Lecture Notes in Computer Science 1390

Edited by G. Goos, J. Hartmanis and J. van Leeuwen

Springer

Berlin
Heidelberg
New York
Barcelona
Budapest
Hong Kong
London
Milan
Paris
Santa Clara
Singapore
Tokyo

Christian Scheideler

Universal Routing Strategies for Interconnection Networks

Springer

Series Editors

Gerhard Goos, Karlsruhe University, Germany
Juris Hartmanis, Cornell University, NY, USA
Jan van Leeuwen, Utrecht University, The Netherlands

Author

Christian Scheideler
Universität-Gh-Paderborn, Fachbereich 17
Mathematik-Informatik und Heinz-Nixdorf Institut
Fürstenallee 11, Gebäude F, D-33102 Paderborn, Germany
E-mail: chrsch@uni-paderborn.de

Cataloging-in-Publication data applied for

Die Deutsche Bibliothek - CIP-Einheitsaufnahme

Scheideler, Christian:
Universal routing strategies for interconnection networks / Christian
Scheideler. - Berlin ; Heidelberg ; New York ; Barcelona ; Budapest ;
Hong Kong ; London ; Milan ; Paris ; Santa Clara ; Singapore ;
Tokyo : Springer, 1998
  (Lecture notes in computer science ; Vol. 1390)
  ISBN 3-540-64505-5

CR Subject Classification (1991): C.2, F.1.1-2, G.2.2, F.2.2, B.4.3

ISSN 0302-9743
ISBN 3-540-64505-5 Springer-Verlag Berlin Heidelberg New York

Typesetting: Camera-ready by author
SPIN 10636879   06/3142 – 5 4 3 2 1 0   Printed on acid-free paper

To my parents
and Gabriele

Foreword

During the last decade, the need for realizing communication in networks has grown enormously. The use of telecommunication networks, local area computer networks, or wide area networks like the Internet is growing rapidly, and the current trend to integrate different types of data like text, audio, or video is further increasing the demand for high bandwidth, low latency communication networks. Similarly, tightly coupled parallel computer systems demand sophisticated communication devices.

The importance of realizing communication in networks has motivated intensive worldwide research activities, also in basic computer science research. Many proposals have been made for the topology of communication networks: meshes, hypercubes, hypercube-like networks such as the butterfly, shuffle exchange, de Bruijn and fat tree, and expanders. Many routing modes have been proposed: circuit switching, store-and-forward routing, wormhole routing, etc. Distinctions between oblivious and adaptive routing have been introduced, and hardware restrictions like a limited buffer size or edge-bandwidth have been taken into consideration. A large collection of routing protocols has been developed and analyzed for the networks and modes mentioned above.

This monograph provides a comprehensive description of the current research on routing. In particular, it presents essential new contributions on universal routing. By introducing the routing number of a network, i.e., the offline routing time for worst case permutations, Christian Scheideler offers a rigorous approach to measure the quality of routing protocols. He applies it to known protocols like Ranade's random rank protocol, its variants for bounded and unbounded buffers, and extensions to arbitrary networks. The main contributions are new universal protocols for store-and-forward and wormhole routing with small buffers and without buffers (deflection routing). He examines the benefits of large edge-bandwidth, and the capabilities and limitations of deterministic protocols and of bounded storage capabilities of the switches.

This monograph is an extension of Christian Scheideler's dissertation (Ph.D. thesis), submitted to Paderborn University in 1996. I am deeply impressed by his intuition about combinatorial and probabilistic phenomena, and by his capabilities to understand, apply, and extend deep combinatorial and probabilistic methods.

November 1997
Friedhelm Meyer auf der Heide
Paderborn University, Germany

Preface

Routing strategies are methods that use the available infrastructure between processing units to enable the exchange of information between them. The main components of routing strategies form methods for the selection of routes in a communication network and methods for scheduling the motion of the messages along their selected routes. Efficient routing strategies are of fundamental importance for two reasons.

We live in an era that is often referred to as the Information Age. Analyzing, manipulating, gathering, and distributing information has become an important economic and therefore political factor. Telecommunication networks, computer networks in companies and universities or the Internet are examples of networks that have been designed for a fast exchange of information and execution of requests (for instance, telephone calls, emails, money transfers between banks). The vast growth of users and services especially in metropolitan and wide-area networks makes it necessary to exploit the available performance of these networks as well as possible.

On the other hand, there are many scientific problems, such as modeling global weather patterns, analyzing the aerodynamic properties of a wing, and simulating the strange subatomic world of quantum theory, that require enormous volumes of computation, which cannot be handled by single processor systems. Hence multiprocessor machines are needed that can efficiently execute parallel programs that serve these purposes. Since the time for one computation step is usually much shorter than the time needed for exchanging information, much work has been invested in recent years to speed up the communication time as much as possible.

Many routing strategies have already been developed for specific classes of network topologies. Of special interest are so-called *universal* routing strategies, that is, strategies that can be applied to arbitrary network topologies. In addition to providing a unified approach to routing in standard networks, the advantage of universal routing strategies is that they are ideally suited for communication in irregular networks that are used in wide-area networks and that arise when standard networks are modified or develop faults. Furthermore, universal routing places no restrictions on the pattern of communication that is being implemented (such as requiring that it form a permutation). Hence these algorithms are ideally suited for any communication problem that may occur during the execution of a parallel algorithm.

Since there have been significant advances in the development and analysis of universal routing strategies only during the last few years, almost none of the results have found their way into textbooks yet. The purpose of this monograph, which is based on my thesis written in 1996 at the Paderborn University, is therefore to present the history and state of the art of universal routing strategies.

The text is self-contained with respect to the historical background and notations used. However, it is recommended that the reader has some experience with using probabilistic arguments in proofs.

Chapter 1 gives an introduction to the area of communication strategies and presents research areas to which routing has a close relationship. In Chapter 2 we motivate our routing models by describing how routing is done in practice. Chapter 3 contains all terminology needed for the following chapters. After giving some basic definitions in probability theory and graph theory, we define a hardware model for parallel systems with point-to-point communication, the routing problem, routing strategies, models for passing messages, and models for storing routing information. The rest of this book gives a survey on universal routing strategies for the two most important message passing models studied in theory: the store-and-forward routing model and the wormhole routing model.

Chapters 4 to 9 deal with store-and-forward routing strategies. Chapter 4 gives a survey of the history of store-and-forward routing protocols. It presents both results about protocols for specific networks and universal routing protocols. Furthermore, networks are presented for which optimal randomized and deterministic store-and-forward routing protocols are already known. Chapter 5 introduces an important parameter for measuring the routing performance of networks. This parameter is called the *routing number*. The nice property of this parameter is that it describes the routing performance of networks more accurately than other parameters that have been used so far, such as the expansion or bisection width of a network. Furthermore, it can be easily used to demonstrate how close the performance of routing protocols can get to the optimal routing performance of networks. In Chapter 6, we prove the existence of efficient offline protocols for routing packets along a fixed path collection and show how these protocols can be applied to network simulations. The proofs for the first two offline protocols concentrate on minimizing the routing time, whereas the proof for the third offline protocol concentrates on minimizing the available space for buffering packets during the routing. Chapter 7 gives an overview of the best universal oblivious protocols for store-and-forward routing known so far. It is shown how each of these protocols can be applied to routing in specific networks, and what the limitations of these protocols are. In Chapter 8, a historical survey of adaptive routing protocols is given. Furthermore, some techniques for developing universal adaptive protocols are presented. One of these techniques is called "routing via simulation." It will be demonstrated how this strategy can be used to construct efficient deterministic routing protocols for arbitrary networks. Chapter 9 gives an overview of compact routing strategies, that is, strategies for networks in which the space for storing packets and routing information is limited. After summarizing results about how space restrictions influence the efficiency of selecting routes, we present trade-offs between space requirements and the time necessary to route messages in arbitrary networks. The "routing via simulation" technique presented in the previous section is generalized to a strategy that yields efficient universal compact routing protocols.

Chapters 10 to 12 deal with wormhole routing strategies. Chapter 10 gives a survey of the history of wormhole routing protocols. It presents both, results about protocols for specific networks, and results about universal routing protocols. Furthermore, upper and lower bounds are shown for wormhole routing in arbitrary networks. In Chapter 11 we describe two universal oblivious wormhole routing protocols and show how they can be applied to arbitrary networks. We further show how to improve the performance of one of these protocols for specific classes of networks such as butterflies and meshes. Chapter 12 deals with all-optical routing. We present nearly tight bounds for the runtime of a very simple protocol for sending messages along an all-optical path collection. We further show how this protocol can be applied to specific networks.

The monograph ends with a summary of its results and important open problems in the field of routing theory. Also, an outlook on future directions will be given.

Large parts of this book have been previously published as joint work. The results of Chapter 5 partially extend results in [MS96b]. The idea behind the third offline protocol in Chapter 6 is based on [MS95b]. Chapter 7 contains results that partially extend [MS95a, CMSV96]. Large parts of Chapters 8 and 9 are extensions of [MS95a, MS96a]. Chapter 11 is mostly based on [CMSV96], and Chapter 12 consists of results in [FS97].

Acknowledgements

First of all, I would like to thank Prof. Dr. Friedhelm Meyer auf der Heide for his great support and many helpful discussions. I am obliged to him for helping me writing the papers that led to this book and for establishing many important national and international contacts that influenced my work. Furthermore, I would like to thank Berthold Vöcking for the very productive collaboration and many valuable discussions. I also want to express my thanks to all members of the research group Meyer auf der Heide for the good cooperation and the nice atmosphere.

Besides my colleagues, I am very grateful to my parents whose continuous support greatly helped me to finish my diploma and this thesis. Last, but not least, I thank my wife Gabriele for her great support and for enduring all the days in which she hardly ever saw me because of my work.

Paderborn, November 1997   *Christian Scheideler*

This work has been partly supported by

- DFG-Sonderforschungsbereich 376 "Massive Parallelität: Algorithmen, Entwurfsmethoden, Anwendungen",
- DFG Leibniz Grant Me872/6-1, and by
- EU ESPRIT Long Term Research Project 20244 (ALCOM-IT)

Table of Contents

1. Introduction

Efficient communication is a prerequisite to exploit the performance of large parallel systems. For this reason much work has been invested in recent years to develop efficient communication mechanisms. This includes the development of hardware designs with different characteristics, as well as the design of efficient communication software. In this book a survey will be given about theoretical results of designing efficient routing strategies for communication in various hardware models. The aim of this monograph is to present the state of the art concerning routing strategies that are universally applicable and to deepen the understanding in how hardware restrictions influence the asymptotic behavior of routing strategies.

To understand the importance of routing, let us give a brief overview of the history of parallel processing, which includes the development of parallel systems (Section 1.1) as well as theoretical research in parallel algorithms and architectures (Section 1.2) and routing strategies (Section 1.3). This is followed by an overview of research areas that have a close relationship to routing (Section 1.4) and a survey of the main contributions of this book (Section 1.5).

1.1 The Emergence of Parallel Systems

Parallel processing is best defined by contrasting it with normal serial processing. At the 1947 Moore School lectures, John von Neumann and his colleagues propounded a basic design, or architecture, for electronic computers in which a single processing unit was connected to a single store of memory, the so-called RAM (*random access machine*). In a RAM, the processor fetches instructions from the memory, performs a calculation, and writes the results back to the memory.

The RAM was popular for several reasons. First, it was conceptually simple: only one thing was going on at a time, and the order of operations corresponded to what a human being would do if he or she were carrying out the same computation. This conceptual simplicity was very important in the early days of computer science, when little was known about how to write a program for an automatic calculation machine.

Second, RAMs were simpler to build than any of the alternatives, since they contained only one of everything. Third, Grosch's Law, which was promulgated at the time, held that the performance of a computer was proportional to the square of its cost. This was because the basic components of computers (vacuum tubes and magnetic drums) were fragile, error-prone devices.

Von Neumann and his colleagues did not ignore the possibility of using many processors together. Indeed, von Neumann was perhaps the originator of the idea of cellular automata, in which a very large number of simple calculators work simultaneously on small parts of a large problem. However, the hardware technology of the time was not capable of creating such machines, and the software technology was not capable of programming them.

With the switch from vacuum tube to solid state components in the 1960s and the development of vector computers and VLSI (*very large scale integration*) technology in the 1970s on the hardware side, and the development of concurrent programming methods by the late 1970s on the software side, parallel computers became realizable. Today, the number of companies producing parallel computers, and the number of different parallel computers being produced, is growing rapidly. There are two arguments that show that the future of high-performance computing belongs to them.

The first is economic. Parallel computers tend to be much more cost effective than their serial counterparts. This is primarily because of the economics of VLSI technology: ten small processors with the same total performance of one large processor almost invariably cost less than that one large processor.

The second argument is based on a fundamental physical law. Because information can not travel faster than the speed of light, the only ways of performing a computation more quickly are to reduce the distance information has to travel, or to move more bits of information at once. Attempts to reduce distance are eventually limited by quantum mechanics – a computer whose wires are single atoms might be physically realizable, but would have to incorporate enormous levels of redundancy to compensate quantum uncertainty. Moving more bits at once is parallelism, and this is the approach which is proving successful. Today, every major player in the supercomputing game is building machines which use several processors together in order to solve a single problem. The "only" questions remaining are how many processors should be used, how powerful should they be, and how should they be organized.

1.2 Theoretical Research in Parallel Computing

Theoretical research in parallel computing began to flourish with the design of a parallel model of the RAM, the PRAM (*parallel random access machine*) [FW78, Go78, SS79] . The PRAM model consists of a fixed (arbitrarily large) set of processors and a single, so-called *shared*, memory. The processors work synchronously and have random access to the shared memory cells. Since the user does not have to worry about synchronization, locality of data, communication capacity, delay effects or memory contention, the PRAM represents an idealization of a parallel computation model.

On the other hand, PRAMs are very unrealistic from a technological point of view; large machines with shared memory can only be built at the cost of very slow shared memory access. Hence research in parallel systems basically follows two directions today: finding efficient simulations of a PRAM on more realistic models, or developing computational models for efficient parallel algorithms on arbitrary parallel systems.

Finding efficient PRAM simulations has been pioneered by Mehlhorn and Vishkin [MV84]. Valiant pioneered the development of computational models for efficient parallel algorithms on arbitrary parallel systems by developing the BSP model [Va90].

Although simulating the PRAM on parallel systems is convenient for the user, even optimal simulations are in general much slower than optimal implementations on these systems. Hence it would be more desirable to have models that take into account issues like synchronization time, locality of data, and delay effects as it is partly done in the BSP model. In order to minimize these effects for the runtime of parallel algorithms, fast routing hardware and routing strategies are highly needed. In recent years, much work has been invested by the research community to find efficient hardware models and routing strategies.

In case of hardware models, two types of parallel systems have been extensively studied: processors that communicate via a bus, and processors that exchange information via point-to-point communication links.

The first type of parallel systems is non-scalable, since a bus is able to forward only a fixed amount of messages at each time step. Hence such systems can only can be used for efficient communication if the number of processors connected to a bus is not too high. In fact, the slowdown of bus systems w.r.t. the communication time is linear to the number of processors. For networks with point-to-point communication, the slowdown can be reduced to grow only logarithmic to the number of processors. Hence, asymptotically, networks with point-to-point communication are much more efficient than bus topologies.

Many communication strategies have already been developed for specific classes of network topologies. Of special interest are so-called *universal* routing strategies, that is, strategies that can be efficiently applied to arbitrary network topologies. In addition to providing a unified approach to routing in standard networks, the advantage of universal routing strategies is that they are ideally suited to routing in irregular networks that are used in wide-area networks and that arise when standard networks are modified or develop faults. Furthermore, universal routing places no restrictions on the pattern of communication that is being implemented (such as requiring that it form a permutation). Hence these protocols are ideally suited for any communication problem that may occur during the execution of a parallel algorithm. This book will therefore concentrate especially on describing universal routing strategies.

1.3 History of Routing

The earliest communication networks established were telephone networks. Important problems at that time were to find architectures and techniques that allow as many lines as possible to be established between users. In theory, these problems were often reduced to the problem of finding a multistage network that allows arbitrary pairwise connections between N input terminals and N output terminals. These networks are often referred to as *nonblocking* and *rearrangeable* networks or connectors. See [Pi82] for an excellent survey and [ALM96] for more comprehensive descriptions of previous results.

Networks in general that preallocate transmission bandwidth for an entire call or session are called *circuit switching networks*. Before 1970, virtually all interactive data communication networks were circuit switched, the same as the telephone network. However, since most interactive data traffic occurs in short bursts, a large percentage of the bandwidth is wasted. Thus, as digital electronics became inexpensive enough, it became dramatically more cost-effective to completely redesign communication networks, introducing the concept of packet switching where the transmission bandwidth is dynamically allocated, permitting many users to share the same transmission line previously required for one user. Packet switching has been so successful, not only in improving the economics of data communications but in enhancing reliability and functional flexibility as well, that in 1980 virtually all new data networks being built throughout the world were based on packet switching. For a survey on the early history of packet switching technology see an article by Roberts [Ro78].

In principle, packet switching is highly flexible to any kind of bit rate or changing bit rates. However, in order to cope with all kinds of traffic patterns, packet switches require strategies for flow control and therefore need a much more sophisticated processing unit than circuit switches. Analyzing strategies for flow control proved to be an extremely difficult task. Early results used queueing theory and could only handle very simple network topologies such as a ring. Therefore most of the early results where obtained by using simulations and field trials. See [GK80] for an early history of flow control models and techniques.

The most simple model developed at that time to analyze flow control strategies was the store-and-forward routing model. In this model time is partitioned into synchronous steps. One step is defined as the time a packet needs to be sent along a link. A node must store the entire packet before it can forward any part of it along the next link. Using this model, Leighton, Maggs and Rao achieved a major breakthrough in 1988 by proving the following result [LMR88].

Consider an arbitrary set of loop-free paths such that the longest path has length D (the *dilation* of the path collection) and at most C paths cross any edge (the *congestion* of the path collection). Then there exists a strategy of sending one packet along each of these paths in time $O(C + D)$, using only constant size buffers.

This result was remarkable, since it only requires two paramaters to specify a routing problem: the dilation and the congestion. Since for any path collection with dilation D and congestion C, at least $\max\{C, D\}$ steps are needed to send one packet along each path, the upper bound above is asymptotically optimal. The only drawback of the result was that its proof is non-constructive. However, in 1996, Leighton, Maggs and Richa [LMR96] could present an algorithm that computes such a schedule in polynomial time.

Besides the result above, Leighton *et al.* could also present in the same paper [LMR88] a simple distributed local control algorithm (or *online* algorithm for short) that routes packets along an arbitrary collection of n loop-free paths with congestion C and dilation D in time $O(C + D \log(nD))$ with high probability. Subsequently, many people tried to improve this result. Major breakthroughs in this area have only been achieved recently. We will present later a detailed survey of the respective results.

Besides the efforts of finding efficient routing algorithms (also called routing protocols) for fixed path collections, research was also done in how to find efficient path collections in networks. One approach was to construct a system of paths, one path for each pair of nodes, in a preprocessing phase and store it in the nodes of the network. Given a routing problem, the packets then simply have to follow the respective paths in this system. However, as shown by Borodin and Hopcroft [BH85], for any path system in any bounded degree network of size N there exists a permutation routing problem that has a congestion of $\Omega(\sqrt{N})$. Since the diameter of these networks can be as low as $O(\log N)$, this congestion bound is unacceptably high. So people also tried to find strategies that ensure that for every permutation routing problem the congestion is low. Meyer auf der Heide and Vöcking [MV95], for instance, showed how to find path collections with low congestion and dilation online for every permutation routing problem in arbitrary node-symmetric networks (which includes many standard networks such as the hypercube, butterfly, and torus).

Another direction was to find routing protocols that do not send their packets along fixed paths, but rather adaptively choose the paths, depending on situations like the current contention at nodes or links. Such protocols are called *adaptive* (in contrast to *oblivious* protocols that use fixed paths). Adaptive protocols have several advantages over oblivious protocols, in particular when routing in faulty networks. However, it proved to be much more difficult to analyze them, which explains why much more results are known about oblivious protocols than about adaptive protocols. Mostly, the development of adaptive protocols has been restricted to specific networks like the mesh (see, e.g., Chinn *et al.* [CLT96]).

The research community dealing with routing problems has been rapidly growing since the last decade. The most important questions investigated can be summarized as follows: How much does adaptive routing improve over oblivious routing? How much does randomness help? How does it help if each node can have a large number of neighbors? What benefit is available if a node can send packets to several neighbors within a single time step? Borodin *et al.* [BRSU93] managed to obtain a hierarchy of time bounds for worst case permutation routing. Their results are summarized in Figure 1.1. In this figure, the letters A and O distinguish adaptive (A) and oblivious (O) routing schemes. The letters R and D distinguish randomized and deterministic schemes. The letters M and S distinguish multi-port and single-port routing. In multi-port routing, a node is allowed to send a packet along each of its outgoing links simultaneously, whereas in single-port routing only one outgoing link may be active at any time. The time bounds displayed represent the worst case permutation routing time for a best possible network of size n and maximum degree d. The edges in the figure show which models are strictly weaker. The ARM model is the strongest, while the ODS model is the weakest.

Note that the upper bounds for ADM and ORM were obtained by using *different* networks. The question therefore remains open, how these models relate to each other for any *specific* network, or whether large classes of networks can be identified (such as the class of planar networks) for which, say, ADM and ORM are asymptotically equal for *any* network within this class (as we will see, this is indeed true for the class of bounded degree, planar networks!).

Since in practice routing chips only have a limited space for storing packets, instructions, and routing tables, many researchers also dealt with the problem of how much influence space restrictions can have on the routing performance of networks. Many networks have been identified that allow the efficient routing of data even under severe hardware restrictions. Pippenger [Pi84], for instance, was the first who could show how to route in the butterfly in optimal time even if only constant size buffers are available. Also strategies have been studied that require no buffer at all. These are usually referred to as *hot potato* routing strategies (see, e.g., Chinn *et al.* [CLT96]). People also studied what effects space restrictions for instructions and tables can have on the routing performance of networks, and especially for the design of routing structures that allow each message to find its way through the network to its destination. A nice survey of results in this area can be found in a paper by Fraigniaud and Gavoille [FG96] (see also Chapter 9).

Besides the store-and-forward routing model, other models were used such as the wormhole routing model. This model owes much of its recent popular-

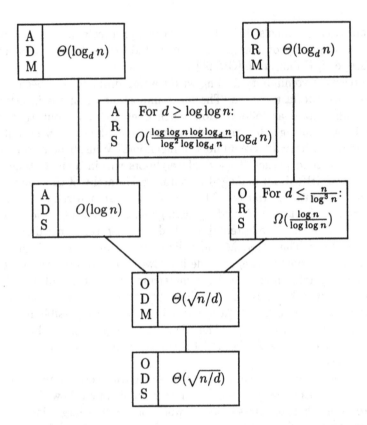

Fig. 1.1. The routing hierarchy.

ity to an influential paper by Dally and Seitz [DS87]. In wormhole routing, messages are sent as *worms*, each of which consists of a sequence of fixed size packets called *flits*. A time step is defined as the time it takes to send a flit along a link. Since only the first flit contains the destination of the worm, in order to avoid confusion at the links the flits of a worm must travel in a contiguous fashion, i.e. like a real worm. This model has several advantages compared to store-and-forward routing we will discuss later in detail. However, performance degradation can be severe in case of heavily loaded networks. In order to reduce the latency and contention, *virtual* channels were introduced. These channels allow a link to be crossed by many worms at the same time. The number of virtual channels supported by a link will be called its *bandwidth* in this book. In addition to the questions above, the question was raised of how much does the link bandwidth of a network influence its routing performance. Partial answers to this question have only been obtained recently. They will be presented in detail in this book.

Routing theory is not only important solely for the purpose of communication, but also can be used for other areas. On the other hand, results from other research areas contributed to advances in the routing area. In the following, we give an overview on some of the most important areas routing has to close relationship with.

1.4 Research Areas Related to Routing

In this section we present areas that use or have some connection to routing. In particular, we give an overview on parallel sorting, shop scheduling, multicommodity flow, network emulations, shared memory simulations, and load balancing. For each of these areas we either show how routing strategies can help to find solutions, or describe what consequences results in this area have for finding efficient routing schemes.

1.4.1 Parallel Sorting

Given n records distributed uniformly over the n processors of some fixed interconnection network, the *sorting problem* is to route the record with the ith largest associated key to processor i, $0 \leq i < n$. Distributed sorting algorithms are of importance for routing, since any sorting algorithm that needs worst case time T can be used to route any permutation in time T by sorting the packets according to their destination address.

One of the earliest parallel sorting algorithms is Batcher's bitonic sort [Ba68], which runs in $O(\log^2 n)$ time on the hypercube, shuffle-exchange, and cube-connected cycles of size n. Leighton [Le85] exhibited a bounded-degree network of size n based on the sorting circuit by Ajtai, Komlós, and Szemerédi [AKS83] that can sort n records in time $O(\log n)$. It is still an open problem whether a sorting time of $O(\log n)$ can be reached by a hypercube and related networks. The best result so far is due to Cypher and Plaxton [CP93]. They presented a deterministic protocol for sorting n records on a hypercube, shuffle-exchange, and cube-connected cycles of size n in time $O(\log n (\log \log n)^2)$.

A variety of sorting algorithms has been proposed for the two-dimensional mesh, starting with the work of Orcutt [Or74] and Thompson and Kung [TK77]. Schnorr and Shamir [SS86] gave a $3n + o(n)$ time sorting algorithm on the $n \times n$ mesh for the case that each processor can only hold a single packet at any time. They further show a nearly matching lower bound of $3n - o(n)$ steps. The fastest known algorithms for the case that every processor can hold a constant number of packets at a time run in $2n + o(n)$ steps [KK92, KSS94].

Early examples of sorting algorithms for multi-dimensional meshes were given by Thompson and Kung [TK77] and Nassimi and Sahni [NS79]. For any constant d, Kunde [Ku91] and Suel [Su94] showed that there is a deterministic sorting algorithm for the d-dimensional mesh that is at most a factor of two away from the lower bound, which implies fast deterministic permutation routing protocols for these networks.

1.4.2 Shop Scheduling

Shop scheduling models are defined in terms of resources, called *machines*, and activities, or *jobs*, each of which consists of a set of *operations*. Each operation of a job has an associated machine that it must be processed on for a given length of time. Each machine can process at most one operation at a time, and each job can have at most one of its operations undergoing processing at any point in time. The problem is to deliver the jobs among the machines in such a way that the time to complete all jobs becomes minimal.

Depending on how (or whether) the order of a job's operations are restricted we obtain three well-studied models: In the *open shop* problem, the operations of a job can be processed in any order, whereas a *job shop*, a job's operations are completely ordered, and the ordering is job-dependent. In a *flow shop*, each job has exactly one operation for each machine, and the ordering of the operations of a job is the same for all jobs; thus the flow shop is a special case of the job shop.

Shop scheduling problems are notorious for their intractability, both in theory and in practice. The first algorithm for constructing schedules for arbitrary job shop problems with polylogarithmic approximation ratio is due to Shmoys, Stein and Wein. In [SSW94] they present a deterministic $O(\frac{\log^2(m\mu)}{\log\log(m\mu)})$-approximation algorithm, where μ is the maximum number of operations per job and m is the number of machines. When m is fixed, they give a polynomial $(2 + \epsilon)$-approximation algorithm for job shops [SSW94]. In case of flow shops, Hall [Ha95] provides a polynomial $(1 + \epsilon)$-approximation algorithm. Among other results, Goldberg *et al.* [GPSS97] were able to improve the approximation ratio of the algorithm by Shmoys *et al.* for the general case by a $\log\log$ factor.

Assume that each job occupies each machine that works on it for a unit of time, and that no machine has to work on any job more than once. If we interpret every job as a packet, and every machine as an edge, then the problem of finding a schedule can be transformed into the problem of sending packets along a suitably chosen simple path collection. Let us define the *dilation* of the scheduling problem to be the maximum number of machines that must work on any job, and the *congestion* to be the maximum number of jobs that have to be run on any machine. Leighton, Maggs, and Rao [LMR94] showed that for any simple path collection with congestion C and dilation D there exists a protocol for routing packets along the paths in time $O(C + D)$. Hence it is possible for the special case described above to construct a constant-factor approximation algorithm for an arbitrary number m of machines.

Note that in a flow shop in which all operations have unit length the scheduling problem is trivially solvable, since the jobs can simply be pipelined through the machines in $n + m - 1$ time units, where n is the number of jobs.

1.4.3 Multicommodity Flow

The multicommodity flow problem involves simultaneously shipping several different commodities from their respective sources to their sinks along edges of a given network so that the total amount of flow going through each edge is no more than its capacity. Associated with each commodity is a demand, which is the amount of that commodity that we wish to ship. Given a multicommodity flow problem, one often wants to know if there is a *feasible flow*, i.e., if it is possible to find a flow which satisfies the demands and obeys the capacity constraints. More generally, we might wish to know the maximum percentage z such that at least z percent of each demand can be shipped without violating the capacity constraints. The latter problem is known as the *concurrent flow problem* and is equivalent to the problem of determining the minimum ratio by which the capacities must be increased in order to ship 100% of each demand.

Let n, m, and k denote the number of nodes, edges, and commodities in the network. The concurrent flow problem can be formulated as a linear program in $O(mk)$ variables and $O(nk+m)$ constraints. Linear programming can be used to solve the multicommodity flow problem optimally in polynomial time (see, e.g., [KV86]). Fast combinatorial approximation algorithms for concurrent flow problems can be found in [KPST90, LMP+95, GVY93, PT93].

Consider multicommodity flow problems in which each source has an outgoing flow of one, and each sink has an incoming flow of one. If we make use of Raghavan's [Ra88] method of converting fractional flows into integral flows, then a feasible fractional flow for edges with capacity C implies that a collection of paths can be chosen to connect all source-sink pairs with congestion $O(C+\log m)$, where m is the number of edges in the network. Hence solutions to multicommodity flow problems can be used to prove the existence of path collections for some given congestion bounds.

1.4.4 Network Emulations

Consider the problem of emulating a network G by a network H. There are three different kinds of strategies.

- **Static 1–1 embedding**: Every processor in G is simulated by one fixed processor of H.
- **Static embedding with redundancy**: Every processor in G is simulated by at least one fixed processor of H.
- **Dynamic embedding**: At every time step, every processor in G is simulated by at least one processor of H.

In order to simulate the traversal of a packet along an edge in G, the respective packet usually has to be sent along a path in H. Hence we need routing strategies for simulating a communication step in G. In case of static embeddings, many results are known. An overview on results in this area has been given by Monien and Sudborough [MS90].

The model of dynamic embeddings was first considered by Meyer auf der Heide [Me83]. There are several results indicating the strength of these simulations, see [KLM+89, Sch90, KKR93b]. E.g., the $\sqrt{n} \times \sqrt{n}$-mesh can be simulated on an n-node butterfly network with constant slowdown [KLM+89], whereas a static embedding has dilation (and therefore slowdown) $\Omega(\log n)$, as shown in [BCH+88].

Let us call a network n-*universal with slowdown* s, if it can simulate T steps of any network of size n with slowdown at most $s \cdot T$ for any $T \geq 1$. Galil and Paul [GP83] show that every network H of size m that can sort n numbers in time $sort(n, m)$ is n-universal with slowdown $O(sort(n, m))$. Hence the sorting algorithm presented by Cypher and Plaxton [CP93] implies that the shuffle-exchange network and the hypercube of size n are n-universal with slowdown $O(\log n \cdot (\log \log n)^2)$. In [Me86], Meyer auf der Heide presents an $n^{1+\epsilon}$-node n-universal network with constant slowdown. Kaklamanis *et al.* [KKR93b] present, among other results, a network of size n that can simulate every planar network of size n with slowdown $O(\log \log n)$. For restricted classes of bounded degree networks (i.e., networks where the size of the t-neighborhood of each node is bounded by a polynomial in t), Meyer auf der Heide and Wanka [MW89] showed that constant slowdown simulations only need $O(n \cdot poly(\log n))$-size universal networks. Lower bounds for dynamic embeddings are shown in [Me83, Me86, KLM+89, KR94, MSW95]. In [MSW95], Meyer auf der Heide *et al.* show that, if H is an arbitrary constant-degree network of size m that can simulate all constant-degree networks of size n with slowdown s, then $m \cdot s = \Omega(n \log m)$.

1.4.5 Shared Memory Simulations

Parallel machines that communicate via a shared memory, so-called *parallel random access machines* (PRAMs), represent an idealization of a parallel computation model. The user does not have to worry about synchronization, locality of data, communication capacity, delay effects or memory contention.

On the other hand, PRAMs are very unrealistic from a technological point of view; large machines with shared memory can only be built at the cost of very slow shared memory access. Therefore there is a need for finding efficient simulations of a PRAM on more realistic models such as bounded degree networks. The main problem is the distribution of the shared memory cells among the nodes of the network to allow fast accesses. A standard method is to use universal hashing (see, e.g., [MV84, KU86, UW87, KLM93, GMR94]). For a survey on high performance hash functions see, e.g., [Di91].

A PRAM consists of processors $P_1, \ldots, P_n$ and a shared memory with cells $U = \{1, \ldots, p\}$, each capable of storing one integer. The processors work synchronously and have random access to the shared memory cells. A PRAM in which no two processors are allowed to access the same shared memory cell at the same time during a read or write step is called EREW (*exclusive-read exclusive-write*) PRAM. A PRAM that allows simultaneous accesses of many processors to one memory cell is called CRCW (*concurrent-read concurrent-write*) PRAM. Of course, if more than one processor wants to write to the same memory cell at the same time, conflicts arise. The following rules are commonly used to resolve these conflicts.

- **Collision rule:** The contents of the memory cell remains unchanged.
- **Arbitrary rule:** An arbitrary processor wins.
- **Priority rule:** The processor with highest number wins.

If we allow *combining* of requests, that is, requests to the same shared memory cell that meet in a node during the routing are combined to one request, Ranade [Ra91] showed that any communication step of a CRCW PRAM of size n can be simulated by a bounded degree network of size n with constant size buffers in optimal time $O(\log n)$, w.h.p.[1] Randomized simulations of PRAMs on complete networks have been presented, e.g., in [MSS96, CMS95]. In [MSS96] Meyer auf der Heide *et al.* show that randomized simulations with double-logarithmic delay exist when using the collision rule for accesses within the complete network, and in [CMS95] Czumaj *et al.* show that randomized simulations with triple-logarithmic delay exist when using the arbitrary rule for accesses within the complete network.

Deterministic simulations of PRAMs have also been studied. Alt *et al.* [AHMP87] show that there is a deterministic algorithm that can simulate any PRAM step on a complete network in $O(\log n)$ steps, and any deterministic simulation on a bounded degree network takes at least $\Omega(\frac{\log n \log p}{\log \log p})$ steps. Herley and Bilardi [HB94] present a protocol that simulates any PRAM step on a suitably chosen bounded degree network in $O(\frac{\log n \log p}{\log \log n})$ steps.

1.4.6 Load Balancing

Consider each processor in a network to have a number of jobs representing its load. The (static) load balancing problem is to distribute the jobs in such a way among the processors, that every processor has (nearly) the same amount of load. Load balancing strategies can be useful for routing problems where at the beginning the distribution of the packets is very imbalanced by first distributing the packets evenly among the processors before they are sent to their destinations. On the other hand, routing strategies might be useful for supporting load balancing strategies in which special, designated processors assign work to other processors.

Load balancing has been studied extensively since it comes up in a wide variety of settings including adaptive mesh partitioning [HT93, Wi91], fine grain functional programming [GH89], job scheduling in operating systems

---

[1] Throughout this book, the terms "*with high probability*" and "*w.h.p.*" mean "with probability at least $1 - n^{-\alpha}$" where $\alpha > 0$ is an arbitrary constant.

[ELZ86, LK87], and distributed game tree searching [KZ88, LuM92]. A number of models have been proposed for load balancing, differing chiefly in the amount of global information used by the algorithm [AAMR93, CK80, Cy89, GM94, LuM93, NXG85].

Local algorithms (that is, algorithms that need no global information) restricted to particular networks have been studied on counting networks [AHS91, KP92], hypercubes [JR92, Pl89], and meshes [HT93, MOW96]. Another class of networks on which load balancing has been studied is the class of expanders. Peleg and Upfal [PU87] pioneered this study by identifying certain small-degree expanders as being suitable for load balancing. Their work has been extended in [BFSU92, He91, PU89a]. These algorithms either use strong expanders to approximately balance the jobs, or the AKS sorting network [AKS83] to perfectly balance the jobs. Thus they do not work on networks with arbitrary topology.

On arbitrary networks, load balancing has been studied under two models. In the first model, any amount of load can be moved across a link in any time step [BT89, Cy89, GM94, HLM+90, SS94]. In the second model, at most unit load can be moved across a link in each time step. Load balancing algorithms for the second model were proposed and analyzed in [AAMR93, GM94, GLM+95, MOW96]. For an overview on efficient load balancing strategies see [Lu96].

1.5 Main Contributions of this Book

Many questions raised in Section 1.3 will be addressed in this book. We will especially concentrate on questions like

- whether, and for which networks, online protocols can reach the best possible time of routing arbitrary permutations,
- whether, and for which networks, adaptive routing protocols are more efficient than protocols using oblivious routing strategies,
- whether, and for which networks, randomized routing strategies are more efficient than deterministic routing strategies,
- how space limitations at the processors (such as bounded buffers or limited space for storing routing tables) or a limited link bandwidth influence the routing time, and
- whether, and under which circumstances, wormhole routing strategies are more efficient than store-and-forward routing strategies for sending messages.

For this, we introduce a parameter called the *routing number* (see Chapter 5). This parameter describes the routing performance of networks very accurately and therefore will be used extensively in the following chapters to apply routing protocols to arbitrary networks. Hence this parameter will be of great importance to answer the questions above.

The class of routing problems we will concentrate on are called *static* routing problems, that is, given a network and a set of source-destination pairs, the task is to route a message for each such pair from the source to the destination. As for the routing problems we always assume that the network given is static in a sense that nodes or links do not develop faults, or connections between nodes do not change. We mainly deal with solving permutation routing problems. The reason for this is that, in order to fully exploit the computational power of a network, it is important that during the execution of a parallel algorithm the work load of the nodes and the communication is as balanced as possible. Since in permutation routing problems every node sends out exactly one packet and receives exactly one packet, these problems are well-balanced and therefore of paramount interest. We further will assume that all nodes operate according to the multi-port model, that is, a node is able to send packets simultaneously along each of its outgoing links. This model has become most common in recent years.

The emphasis of this book will lie on presenting universal routing strategies for both the store-and-forward routing model (Chapters 4 to 9) and the wormhole routing model (Chapters 10 to 12). It will be demonstrated that randomized online routing protocols can be constructed for *any* network that (nearly) reach the optimal worst case time of routing arbitrary permutations (see Chapters 5 and 7). Furthermore, we show that for a large class of networks (which includes all planar networks) even deterministic online routing protocols can be constructed that reach the optimal worst case time of routing arbitrary permutations (see Chapter 8). In addition to this, we present protocols that remain efficient for arbitrary networks even under severe space restrictions (such as limitations on the buffer size or space for storing routing tables, see Chapters 7 and 9). Most of these results will be trade-offs between the runtime the protocol can reach and the space restrictions imposed on the nodes. Furthermore, we show that an increase in the bandwidth (i.e., the number of virtual channels) of the links can significantly improve the performance of some simple routing strategies (see Chapters 7 and 11).

All results mentioned above are presented in detail at the end of the respective chapters and in Chapter 13, which also contains an outlook on future directions. Let us start with describing how routing is done in practice. This will help to motivate our models and terminology in Chapter 3.

2. Communication Mechanisms Used in Practice

Efficient communication has become more and more important in recent years for two reasons.

We live in an era that is often referred to as the *Information Age*. Analyzing, manipulating, gathering, and distributing information has become an important economic and therefore political factor. Telecommunication networks, computer networks in companies and universities or the Internet are examples of networks that have been designed for a fast exchange of information and execution of requests (e.g., telephone calls, emails, money transfers between banks). The main problem for communication networks as described above is to choose communication strategies that ensure some given quality of service requirements and an optimal exploitation of the communication performance of the network.

On the other hand, there are many scientific problems, such as modeling global weather patterns, analyzing the aerodynamic properties of a wing, and simulating the strange sub-atomic world of quantum theory, that require enormous volumes of computation, which can not be handled by single processor systems. Hence multiprocessor machines are needed that can efficiently execute parallel programs that serve these purposes. The main problem for constructing efficient multiprocessor machines is to design architectures and protocols that can ensure a fast and reliable communication between any pair of processors.

In this chapter we give an introduction to methods used in practice to communicate in parallel systems. One technique to allow efficient communication is called *multiplexing*.

2.1 Multiplexing

In communications, *multiplexing* is the process of combining two or more signals and transmitting them over a single transmission link. *Demultiplexing* is the reverse process of separating the multiplexed signals at the receiving end of the transmission link. The number of signals that can be multiplexed together is directly related to available *space*, *time*, and *bandwidth*. Wideband transmission media such as coaxial cable, microwave, and fiber-optic links, for example, have the capacity to multiplex several thousand signals together. There are three fundamental classifications of multiplexing:

- Space-division multiplexing (SDM)
- Wavelength-division multiplexing (WDM)
- Time-division multiplexing (TDM)

Space-division multiplexing (SDM) simply means that different signals are sent along different physical channels. Since a large number of physical channels is expensive and therefore not always available, space-division multiplexing is often combined with time-division multiplexing and/or wavelength-division multiplexing.

Wavelength-division multiplexing utilizes a link's available bandwidth to send multiple signals simultaneously using different wavelengths.

Time-division multiplexing is a method of interleaving, in the time domain, signals belonging to different transmissions. A major difference between TDM and WDM is that WDM is an analog process, whereas TDM is a digital process.

Multiplexing has become an essential part of today's communication systems. It is necessary for us to understand the concept of multiplexing and how it results in an efficient use of the communication links.

In parallel systems with point-to-point communication, messages are directed to their destinations with the help of switching elements.

2.2 Switching Elements

There are basically two types of switching techniques used to forward messages.

2.2.1 Circuit Switches

The purpose of circuit switches is to establish a *circuit* (i.e., a path) between two points of a network for the complete duration of the connection. For each link used by the circuit, this can either be done by reserving a (physical of virtual) channel, or reserving a time slot by using TDM techniques. The latter technique is referred to as STM (*Synchronous Transfer Mode*). In STM, several connections are time multiplexed over one link by joining together several time slots in a *frame*, which is repeated with a certain frequency. A connection will always use the same time slot in the frame during the complete duration of the session. The switching of a circuit of an incoming link to an outgoing link is controlled by a *translation table* which contains the relation of the incoming link and the slot number, to the outgoing link and the associated slot number.

Circuit switching guarantees a fixed bit rate once a connection is established, but on the other hand is very inflexible in supporting different bit rates or adapting to changing bit rates of a connection. Hence circuit switching is not suitable for public or private communication networks that have to support all kinds of services like B-ISDN networks.

2.2.2 Packet Switches

In packet switching, user information is encapsuled in *packets* which contain additional information (in the *header*) used inside the network for routing, error correction, flow control, etc. This transfer mode is called PTM (*Packet Transfer Mode*).

Unlike circuit switches, packet switches need to examine each packet to find out which call it relates to or what its destination is before they can take a routing decision. Early packet switch designs had a typical computer architecture and were controlled wholly by software. With such designs it proved to be very difficult to achieve high levels of traffic handling capacity, and therefore such types of switches can only be used in those data networks where the traffic levels are comparatively low. With the increase in data traffic the reading of packet headers and routing has been carried out more and more by special-purpose hardware or dedicated microprocessors. Packet switches usually contain memory to buffer the packets which are to be transmitted onwards. The load of the buffers varies with the traffic load, giving a variable delay characteristic and eventually a loss of packets in case of full buffers.

In principle, packet switches are highly flexible to any kind of bit rate or changing bit rates. However, in order to cope with all kinds of traffic patterns, packet switches require strategies for flow control and therefore may need a much more sophisticated processing unit than circuit switches.

There are three techniques that have been widely used for the control of switching elements. These are

- **Wired logic**: The call processing decisions and actions are determined by configurations of logic and memory circuits. Such systems make extensive use of logic gates in circuits that are especially designed for the needs of the system. The advantage of wired logic is that it is extremely fast in controlling the switching. The big disadvantage of wired logic is its inflexibility.
- **Action Translator**: The sequences of possible system actions are stored in memory while the call processing is performed in logic circuits specific to the system. Macroinstructions are stored in memory and executed by the logic. "Establish a network connection" might be an example of a macroinstruction for which the logic circuits have been designed. Microprocessors may be used to implement the logic.

- **Stored Program Control (SPC)**: System decisions and actions are taken by a general-purpose processor by reference to a sequence of instructions stored in the memory.

Today, digital switches usually operate under Stored Program Control (SPC). Using SPC, all the necessary operational instructions are held in programmable read only memories (PROMs). Due to the high degree of flexibility that this provides, systems of widely differing size and architecture can be made to operate together in a compatible manner. The principal motives for this development further include:

- the ability to reconfigure the systems to meet developing needs,
- the ease with which new subscribers and subscribers services can be added, and
- improved network management and control of fault conditions.

The connection of processors or computers to powerful parallel systems has been performed in three different levels.

2.3 Local Area Networks (LAN)

Local area networks form the basis for the connection of personal computers and workstations, and are used in all areas of scientific and commercial computing. The most commonly used transmission systems for LANs are the Ethernet and FDDI.

2.3.1 The Ethernet

The Ethernet has dominated the LAN scene over the years and has become an accepted standard. It was developed in 1980 by the combined efforts of Xerox Corporation, Digital Equipment Corporation, and Intel Corporation, and has been accepted under the IEEE 802.3 specification.

The Ethernet is based on a bus. Thick, double-shielded, 50 Ω coaxial cable with a solid center conductor is specified as the transmission medium for Ethernet. Each coaxial cable within a network is referred to as a *segment*. A segment can span a maximum of 500 m. Since nodes must be separated by a minimum of 2.5 m, a segment can connect up to 200 nodes. The raw data rate is 10 Mbit/s, and each node on the segment is driven by a separate 20 MHz clock. The Manchester code format is used to ensure that the clocks are synchronized to the data stream. This is achieved by having at least one transition in each bit interval, regardless of the data bit pattern. A low-to-high transition represents a logic 1, and a high-to-low transition represents a logic 0. The access control protocol used for the Ethernet is called CSMA/CD (*carrier sense multiple access with collision detection*). All nodes listen continually to the network to detect a suitable time to transmit data. If two or more stations try to transmit at the same time, they will continue to transmit for a further 80 ns so that all nodes can recognize that a data collision has occurred. The transmitting nodes then back-off for a random period before trying to retransmit. If a further collision occurs the respective nodes back-off for a longer period. After 16 attempts without success, a node logs a system failure. The transmission frame format consists of a maximum of 1526 bytes and a minimum of 72 bytes. 26 bytes in the frame are reserved for frame synchronization, error detection, and the source and destination address. The data field is variable between the limits of 46 and 1500 bytes.

2.3.2 The Fibre Distributed Data Interface (FDDI)

This network uses optical fibre as the transmission carrier, and operates similar to a *token ring*. A *token* is a special bit pattern or packet that circulates around the ring from node to node. A node desiring the use of the ring for communication takes possession of the token and holds it when it arrives. The node then transmits its packet onto the ring. The packet circulates to its destination node on the ring and returns to the transmitting node, where it is then verified that it was received. The token is afterwards passed forward to the next node.

Unlike normal token rings, an FDDI ring node releases the token immediately after it has transmitted its data. This utilizes the ring bandwidth and time more effectively so that many messages can be circulating simultaneously. The ANSI standard for FDDI specifies a 100 Mbit/s data rate with a distance of up to 2 km between stations. Rings of greater than 100 km circumference can be constructed, and with up to 500 nodes per ring. The network consists of two separate fibres which are connected to each node. This allows the network to be reconfigured under fault conditions at any node. A 4B/5B coding system is used to aid synchronization, that is, each 4-bit symbol is coded into five bits. Of the 32 bit patterns available, 10 are discarded as having too few data transitions, six are used for control purposes and the remaining 16 represent the original symbol. Expanding the data from four bits to five requires a bit rate of 125 Mbit/s. Note that this coding is less rich in transitions than the Manchester code. The Manchester code, however, would require a bit rate of 200 Mbit/s for a data rate of 100 Mbit/s.

The FDDI-2 system has been designed to expand the flexibility of the concept and cater for both packet and circuit switched communications. It supports the 64 Kbit/s circuit switched voice channels of an ISDN (*integrated services digital network*) system. This is achieved by using TDM. Of the 100 Mbit/s rate, 98.304 Mbit/s is available to share between packet and circuit switched applications. The remaining 1.696 Mbit/s is used for synchronization, header and cycle delimiting.

The development of communication networks on the basis of the ATM standard (*asynchronous transfer mode*, see below) influences massively the development of new LANs. Here we can expect that with this technology the number of processors connected to one network can be scaled higher. Moreover the latency is reduced with the help of a better connection and a lower work for injecting a message into the network.

2.4 Wide Area Networks (WAN)

Wide area networks are currently object of vivid discussion in the scientific and public world (consider, e.g., the discussion about broadband ISDN). In general, wide area networks are, like LANs, networks that connect local networks with each other. One of the most popular WANs is the Internet. The following physical transfer modes will most probably be used in future WANs.

2.4.1 The Synchronous Digital Hierarchy (SDH)

Different synchronous multiplex systems have been developed in the past for trials in the USA, UK and Japan. Then, the newly-formed Bellcore specification authority in the USA defined a standard known as SONET (*synchronous optical network*) which replaced an earlier and less flexible synchronous multiplexing proposal known as SYNTRAN (*synchronous transmission*).

SONET was initially compatible only with the North American hierarchy of bit rates. However, in 1988 it was modified in conjunction with the CCITT (*International Consultative Committee on Telegraph and Telephone*, now known as ITU-T) in order to provide options compatible with the European hierarchy (see CCITT recommendations G.707–G.709), so that intercontinental communication is possible. The outcome was the *synchronous digital hierarchy* (SDH). The preferred European options were defined by the European Telecommunications Standards Institute (ETSI) and those for the USA by the American National Standards Institute (ANSI) (see ANSI T1.105-1988 and ANSI T1.106-1988).

SDH covers a wide variety of multiplexing strategies. The basic rate of SDH is 155.520 Mbit/s (in the literature it is often abbreviated to "155 Mbit/s"). This rate is called STM-1, where STM stands for *Synchronous Transport Module*, and at higher rates it is called STM-n, where the rate is n times the basic rate. The STM-n frame structure is an array that consists of 9 rows and $270 \cdot n$ columns of bytes. The order of transmission is left to right then top to bottom. The first nine columns are mostly for *section overheads* (SOH) which provide functions including identification, framing, protection, switching information, error checking, and an embedded operations channel operating at 786 kbit/s. The remaining 261 columns comprise the *payload* into which a variety of signals can be mapped. Each STM-n frame only exists during a point-to-point communication in a network. An STM-n frame is repeated every 125 μs (9 · 270 bytes per 125 μs yields the 155.520 Mbit/s bit rate for the STM-1 frame).

SONET and SDH are expected to be incorporated into existing networks where they will displace existing point-to-point implementations. A number of commercial products are currently available, or have been announced, supporting the SONET or SDH standard.

The virtual transfer mode used in the future will most probably be ATM.

2.4.2 The Asynchronous Transfer Mode (ATM)

Conventional networks carry data in a synchronous manner and because empty slots can be circulating even when the link is needed, network capacity is wasted. This is avoided by the ATM concept, which has been developed for use in broadband metropolitan area networks (MAN) and optical fibre based systems. ATM is supported by both CCITT and ANSI standards, and can also be interfaced to SONET as physical transport layer. Common standards definitions are provided for both private and public networks so that ATM systems can be interfaced to either or both.

With ATM the information carried is broken down into fixed-length cells, the number of such cells per unit time reflecting the bandwidth requirement of the user, and variations in their arrival time reflecting the bursty nature of the traffic. Each cell of 53 bytes is divided into a header of 5 bytes and an information field of 48 bytes. The primary purpose of the header is to identify the connection number for the sequence of cells that constitute a virtual channel for a specific call. The header also contains header error control, generic flow control, and cell loss priority.

The virtual channel a cell belongs to is identified by a *Virtual Channel Identifier* (VCI) of 16 bits stored in the header of that cell. Several virtual channels may be combined into one virtual path while using the same links. The virtual path a cell belongs to is identified by a *Virtual Path Identifier* (VPI) of 8 bits stored in its header. A number of virtual paths may be multiplexed into the same physical path. CCITT defines in Recommendation I.113 how to switch virtual channels and virtual paths. For a more detailed description of ATM see, e.g., [Bl95].

Note that ATM is a transmission system that may be used across a network or subnetwork that contains switching functions. In contrast, SONET is a transmission system for individual links. CCITT is developing a layered approach to the definition of transmission systems in order to define more clearly their interaction and the way they provide transmission services to particular network types.

Next generation LANs and WANs are expected to make increasing use of fiber optics in combination with ATM for digital access and switching protocols, and SONET for link transport (Bellcore, e.g., is already working with a multi-Gbit/s exploratory optical network using ATM/SONET [IL95]). A large number of vendors have already announced their support for this direction in fiber-optic systems development. Additionally, government agencies are recognizing the importance of an information infrastructure for their future competitiveness. In the United States, agencies such as the National Science Foundation (NSF), Advanced Projects Research Agency (ARPA), and the National Institute for Standards and Technology (NIST) have begun to support research aimed for future broadband networks. Special projects such as the National Research and Education Network (NREN) and planned upgrades to the existing electronic mail facilities of the Internet are just part of the efforts to develop a national information infrastructure (NII), the so-called "information superhighway" which is intended to become a nation-wide data communication network. Similar projects are underway in both Europe (the RACE program [Ch94]) and Japan.

One of the critical areas for future development is higher data rate channels. Because of the strong increases in traffic, 2.5 Gbit/s systems have already been deployed (STM-16) by telecommunication companies for interoffice and long haul links. Field trials are currently being conducted on the next step in the digital hierarchy, 10 Gbit/s systems (STM-64). In research laboratories, already systems with 100 Gbit/s capacity and beyond are being explored today. Operation of high speed channels, however, requires high speed electronics to support multiplexing, decision circuits, and other signal processing functions. Given recent prototype demonstrations, there seems to be no reason why these technologies can not be extended to more than 100 Gbit/s. In the long run, however, there will be an increasing price to pay for the further development of high-speed electronic components. Hence they might be the bottleneck for constructing ever faster communication networks.

One solution could be to build so-called *all-optical* networks, that is, networks that maintain the signal in optical form, thereby avoiding the prohibitive overhead of conversion to and from the electrical form. Several research projects are currently exploring the potential of all-optical networks (see [JLT93] for a collection of research proposals). The two basic techniques used in all-optical networks are wavelength division multiplexing and time division multiplexing (see [IEEE94]). The WDM approach allows considerable protocol transparency and bandwidth-on-demand capability, because once a connection is set up at one wavelength, any bit rate framing convention or higher-level protocol stack may be used between node pairs, independent of what is being used in other connections.

Optical technology is not as mature as electronic technology. There are limits to how sophisticated optical processing at each node can be done. In an attempt to keep the processing inside the all-optical network simple, most of the work on all-optical networks has been limited to so-called *broadcast and select networks* until recently. In broadcast and select networks all nodes are connected to one (or more) so-called passive optical star which broadcasts the signals it receives from the nodes to all other nodes. This technique, of course, can only be used for efficient communication if the available bandwidth equals the number of nodes in the system and is therefore not scalable.

Exploratory all-optical networks using all-optical circuit switches are under development. The main problems that have to be solved to construct scalable networks with an efficient and flexible all-optical communication are the development of all-optical switches and filters that allow a fast reconfiguration, all-optical devices for wavelength conversion, and all-optical storage devices.

2.5 Communication Systems for Parallel Computers

Parallel computers are monolithic systems that consist of a number of processors that are connected by some communication network. Such systems have been developed since the middle of the 80s by a number of companies. Examples are the T3D and T3E by Cray, Intel's Paragon, IBM's SP2 and Transputer-based machines by Inmos which were realized by Parsytec and other companies. In contrast to LANs and WANs the processors are very closely connected to each other.

2.5.1 Non-scalable Parallel Computers

Non-scalable parallel computers use as interconnection medium a bus or a physically shared memory.

A system with a physically shared memory that is based on standard processors has been realized in the Alliant FX/2800 [TW91]. This system connects up to seven nodes, each containing four standard processors, via a cache hierarchy with a physically shared memory.

Examples of systems with physical shared memory and specially designed processing units are the Convex-240 and the Cray X-MP, Y-MP, C90, and J90. All systems are based on one chip set. The access to the physical shared memory is especially in all Cray systems very sophisticated. For more details see [TW91].

Parallel computers based on buses are offered today by nearly every computer manufacturer. The communication performance of the buses mainly determines the overall performance of the system. In combination with hierarchical caching schemes, only a small number of processors can be interconnected in this way. Typical parallel computers that use a bus contain up to 32 processors. Systems of this kind that are successful on the market are the Silicon Power Challenge and the SUN Sparc Center systems.

2.5.2 Scalable Parallel Computers

Scalable parallel computers use a network to connect the processors. The network can connect a theoretically unbounded number of nodes that may represent one processor or a complex, non-scalable parallel system. The largest scalable parallel systems that are realizable today contain up to 3000 powerful processors. For very simple processors, systems have been realized with up to 2^{16} processing elements (Thinking Machine CM-2, Marpar MP1). Other examples of scalable parallel systems are the Intel iWarp, Parsytec GC, Intel Paragon, and Cray T3D.

If one considers the milestones in the development of scalable parallel computers, one can see that to a more and more extent standardized processors (that is, processors that can also be found in sequential systems) are used. Intel, for example, uses i860 processors for its Paragon machine, Parsytec offers networks with PowerPC processors, and Cray uses DEC-Alpha processors for its T3D. On the other hand one can see from the entry of established computer companies and the simultaneous vanishing of many pioneer companies in recent years that a consolidation of the market has been taken part, which can also be found in many other areas of rapid development.

Altogether it can be observed that scalable parallel systems are established today in the product palette of large computer manufacturers. This is mainly because it is a commonly accepted fact today that only scalable parallel computers, using standardized processors, will be capable of solving the "grand challenges" with a good price-performance ratio.

The routing of messages in networks is usually controlled by programs called *routing protocols*. In order to avoid the complete reprogramming of routing protocols each time the technology changes, a layered model for software design called the OSI model has been developed.

2.6 The OSI Model

With the ever-increasing need for standards and dependence on standards, the combined efforts of ANSI, CCITT, EIA, IEEE, ISO, and others have led to the development of a hierarchy of protocols called the *Open Systems Interconnect* (OSI) model. The model encourages an open system by serving as a structural guideline for exchanging information between computers, terminals, and networks. The OSI model categorizes data communications protocols into seven levels. The hierarchy of each level is based on a layered concept. Each layer serves a defined function in the network. It depends on the lower adjacent layer's functional interaction with the network. If level 1, the physical layer, for example, were to experience complications, all layers above would be affected. On the other hand, since each layer serves a defined function, that function may be implemented in more than one way. In other words, more than one protocol can serve the function of a layer, thus offering the advantage of flexibility.

Physical Layer. The lowest layer of the OSI model defines the electrical and mechanical rules governing how data are transmitted and received from one point to another. Definitions such as maximum and minimum voltage and current levels are made on this level. Circuit impedances are also defined in the physical layer. An example would be the RS-232-C serial interface specification.

Data Link Layer. This layer defines the mechanism by which data are transported along a link or bus in order to achieve error-free communication. This includes error control, formatting, framing, and sequencing of the data. An example that covers the physical layer and data link layer is the Ethernet.

Network Layer. This layer defines the mechanism by which messages are broken into data packets and routed from a sending node to a receiving node within a communication network. Functions of route selection, flow control, and congestion control are included in this layer for networks of point-to-point links.

Transport Layer. The transport layer is concerned with the establishment and management of end-to-end logical connections between user nodes as well as with improving the quality and characteristics of the communication service provided by the network layer (e.g., functions of end-to-end error control, in-sequence delivery, and reassembly).

The transport layer is the highest layer in terms of communications. Above the transport layer are the session layer, presentation layer, and applications layer. Since these layers are no longer concerned with the technological aspect of the network, we do not present them here in detail.

The OSI model is very famous and used with great success to model all sorts of communication systems. A similar hierarchical architecture as used in OSI is used for the ATM B-ISDN network in CCITT Recommendation I.321 (see Figure 2.1). However, only the lowest layers are explained.

CCITT has not yet determined the relation between ATM/SONET and OSI. The SONET or SDH standard seems to fit best to the data link layer, whereas the ATM standard seems to fall into the network layer.

Although the OSI model has helped a lot in developing standardized and portable communication software, its great disadvantage is that it creates a large communication overhead. In today's computer networks it can happen in the worst case that first messages are split according to the TCP/IP standard, then according to the ATM standard, and finally according to the SONET or SDH standard before they are sent out. This communication overhead implies that fine-grained parallelism seems to be difficult to handle. In-

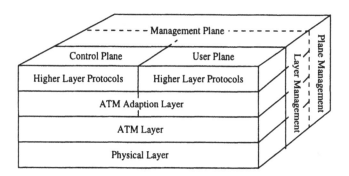

Fig. 2.1. The B-ISDN ATM Protocol Reference Model.

stead, communication should be performed blockwise in order to hide latency and communication overhead. Models that take into account this problem for the development of parallel algorithms are, for instance, the LogGP model [AISS95] and the BSP* model [BDM95].

3. Terminology

In this chapter we present all models and terminology we will need in the following. We start with giving basic definitions and inequalities, followed by a short introduction in probability theory and graph theory. Afterwards, we present in Section 3.4 all models and terminology in the field of routing theory that will be used in this book.

3.1 Basic Definitions and Inequalities

We frequently use the following notations.

- If we use "e" in formulas, we always mean the Euler number (2.71828..).
- By "log" we mean the logarithm to the base of 2. "ln" denotes the logarithm to the base of e. If we use any other base, say b, we write "$\log_b$".
- $\log^k n$ means $(\log n)^k$.
- $\mathbb{N}$ means the set of integers $\{1, 2, 3, 4, \ldots\}$, and $\mathbb{N}_0 = \mathbb{N} \cup \{0\}$.
- By "with high probability" (or "w.h.p." for short) we mean a probability of at least $1 - 1/n^k$ for any constant $k > 0$, where n is the number of elementary events (usually the number of messages) in a random experiment.

Let M be an arbitrary set. Given any subset $A \subseteq M$, $|A|$ denotes the number of elements in A, and $\bar{A} = M \setminus A$. $\mathcal{P}(M)$ is called *power set* of M and consists of all subsets $A \subseteq M$. For any $c \in \mathbb{N}$, the set $[c]$ represents the numbers $\{0, \ldots, c - 1\}$.

We frequently use the following basic inequalities:
For all $n \in \mathbb{N}$,
$$\left(\frac{n}{e}\right)^n \leq n! \leq n^n$$
and
$$\left(1 + \frac{1}{n}\right)^n \leq e \leq \left(1 + \frac{1}{n}\right)^{n+1} .$$
For all $k, n \in \mathbb{N}$ with $k \leq n$ it holds that
$$\left(\frac{n}{k}\right)^k \leq \binom{n}{k} \leq \left(\frac{en}{k}\right)^k \qquad \text{and} \qquad \binom{n}{k} \leq \tfrac{1}{2} \cdot 2^n .$$

3.2 Basic Definitions in Probability Theory

We will use many concepts from probability theory to analyze our protocols and embedding strategies. In this section we will give a short introduction to this theory.

Let Ω be an arbitrary nonempty set. Any element $\omega \in \Omega$ is called *(elementary) event*. We call a function $p : \mathcal{P}(\Omega) \to [0,1]$ *probability measure* on Ω if

- $p(\emptyset) = 0$ and $p(\omega) \geq 0$ for all $\omega \in \Omega$,
- for any subset $\{\omega_1, \ldots, \omega_n\} \subseteq \Omega$ we have $p(\bigcup_{i=1}^n \omega_i) = \sum_{i=1}^n p(\omega_i)$, and
- $p(\Omega) = 1$

In this case we call (Ω, p) *probability space*. A function $X : \Omega \to \mathbb{R}$ is called *random variable* on Ω. X is called *non-negative* if $X(\omega) \geq 0$ for all $\omega \in \Omega$. For the special case that X maps elements in Ω to $\{0, 1\}$, X is called *binary* random variable. The most important measure used in combination with random variables is the expectation.

Definition 3.2.1 (Expectation). *Let (Ω, p) denote an arbitrary probability space and $X : \Omega \to \mathbb{R}$ be an arbitrary function. Then the* expectation *of X is defined as*

$$\mathrm{E}[X] = \sum_{\omega \in \Omega} X(\omega) \cdot p(\omega) \, .$$

The following nice inequality can be shown for arbitrary random variables.

Theorem 3.2.1 (Markov Inequality). *Let X be an arbitrary non-negative random variable with expectation μ. Then, for every $k > 0$,*

$$Pr[X \geq k] \leq \frac{\mu}{k} \, .$$

Of special interest are binary random variables. We distinguish between the following properties of these variables.

Definition 3.2.2. *Given n binary random variables $X_1, \ldots, X_n$, we call them*

- negatively correlated *if for all $i, k \in \{1, \ldots, n\}$, and for all $\{j_1, \ldots, j_k\} \subseteq \{1, \ldots, n\} \setminus \{i\}$*

$$Pr[X_i = 1 \mid X_{j_1} = \ldots = X_{j_k} = 1] \leq Pr[X_i = 1] \, .$$

- positively correlated *if for all $i, k \in \{1, \ldots, n\}$, and for all $\{j_1, \ldots, j_k\} \subseteq \{1, \ldots, n\} \setminus \{i\}$*

$$Pr[X_i = 1 \mid X_{j_1} = \ldots = X_{j_k} = 1] \geq Pr[X_i = 1] \, .$$

In case that equality holds, we call the random variables independent.

Usually, we want to prove results that hold with high probability. An important tool will be the following theorem. It is a combination of results in [AV79] and [Ra88]. See also [Ho87].

Theorem 3.2.2. *Let $X_1, \ldots, X_n$ be n independent binary random variables. Let $X = \sum_{i=1}^{n} X_i$ and $\mu_- \leq \mathrm{E}[X] \leq \mu_+$ for some $\mu_-, \mu_+ \geq 0$. Then it holds for all $\epsilon \geq 0$ that*

$$\Pr[X \geq (1+\epsilon)\mu_+] \quad \leq \quad \left(\frac{e^\epsilon}{(1+\epsilon)^{1+\epsilon}} \right)^{\mu_+}. \tag{3.1}$$

In case that $0 \leq \epsilon \leq 1$, a more careful estimation yields

$$\Pr[X \geq (1+\epsilon)\mu_+] \leq e^{-\epsilon^2 \mu_+/2},$$

and in case that $\epsilon \geq 1$ we can reduce (3.1) to

$$\Pr[X \geq (1+\epsilon)\mu_+] \leq e^{-\epsilon \mu_+/3}.$$

Furthermore, it holds for all $0 \leq \epsilon \leq 1$ that

$$\Pr[X \leq (1-\epsilon)\mu_-] \quad \leq \quad e^{-\epsilon^2 \mu_-/2} \leq \left(\frac{e^\epsilon}{(1+\epsilon)^{1+\epsilon}} \right)^{\mu_-}. \tag{3.2}$$

The bounds in the theorem are usually called *Chernoff-Hoeffding bounds* or *Chernoff bounds* for short. For an elementary proof see [HR90]. Note that Inequality (3.1) also holds if the X_i are negatively correlated, and Inequality (3.2) also holds if the X_i are positively correlated.

3.3 Basic Definitions in Graph Theory

In this section we introduce the basic concepts of graph theory.

A *graph* $G = (V, E)$ consists of a set of *nodes* V and a set of *edges* E. The *size* of G is defined as the number of nodes it contains. For all $v, w \in V$, let (v, w) denote a *directed* edge from v to w, and $\{v, w\}$ denote an *undirected* edge connecting v and w. G is called *undirected* if $E \subseteq \{\{v, w\} \mid v, w \in V\}$, and *directed* if $E \subseteq \{(v, w) \mid v, w \in V\}$. Unless explicitly mentioned, we assume in this book that G is undirected.

A sequence of contiguous edges in G is called a *path*. The *length* of a path is defined as the number of edges it contains. A path is called *node-simple* if it visits every node in G at most once. Similarly, it is called *edge-simple* (or *simple* for short) if it contains every edge in G at most once. G is called *connected* if, for any pair of nodes $v, w \in V$, there is a path in G from v to w. We call a path *cycle* if it starts and ends at the same node. The *girth* of a graph G is defined as the length of the shortest cycle G contains. G is called *tree* if it is connected and contains no cycle. A graph $T = (V', E')$ is called *spanning tree* of G if $V' = V$, $E' \subseteq E$, and T is a tree. G is called *bipartite* if its node set can be partitioned into two node sets V_1 and V_2 such that $E \subseteq \{\{v, w\} \mid v \in V_1,\ w \in V_2\}$.

For any pair of nodes $v, w \in V$, let $\delta(v, w)$ denote the *distance* of v and w in G, that is, the length of a shortest path from v to w. The *diameter* D of G is defined as $\max\{\delta(v, w) \mid v, w \in V\}$. If $\{v, w\} \in E$ then v is called *neighbor* of w. The number of neighbors of v is called the *degree* of v and denoted by d_v. The degree of G is defined as $d = \max\{d_v \mid v \in V\}$. If all nodes in G have the same degree then G is called *regular*.

A family of graphs $\mathcal{G} = \{G_n \mid n \in \mathbb{N}\}$ has degree d if for all $n \in \mathbb{N}$ the degree of G_n is $d(n)$. We are mainly interested in classes of graphs with constant degree. If it is clear to which family a graph belongs we say that this graph has constant (or bounded) degree iff its family has constant degree.

For any subset $U \subseteq V$ we define

$$\Gamma(U) = \{v \in V \setminus U \mid \exists u \in U : \{u, v\} \in E\}\ .$$

G is said to have a *matching* of size k if and only if it contains k node-disjoint edges. If a graph of size N contains a matching of size $N/2$ it is said to contain a *perfect matching*. The following result can be shown about matchings in bipartite graphs. It's proof is similar to the proof of the well-known Matching Theorem by Hall (see [Le92], p. 191).

Theorem 3.3.1. *Let $G = (V_1, V_2, E)$ be an arbitrary bipartite graph with $|V_1| \leq |V_2|$. Then G contains a matching of size $|V_1|$ if for any subset $S \subseteq V_1$ it holds that $|\Gamma(S)| \geq |S|$.*

The following result can be easily shown.

Theorem 3.3.2. *For any graph G with degree d, $d+1$ colors suffice to color the nodes of G in such a way that no two neighbors in G have the same color.*

Proof. Suppose on the contrary that this is not possible. Then there must exist a node in G whose neighbors already use all $d + 1$ colors. Since the degree of any node can be at most d, this can not happen.   □

3.4 Basic Definitions in Routing Theory

In this section we present all parts of a model we need for studying the performance of routing algorithms: the hardware model, the routing problem, routing strategies, message passing models, and a model for storing routing information.

3.4.1 The Hardware Model

We model the topology of a network as an undirected graph $G = (V, E)$. V represents the computers or processors, and E represents the communication links. We assume the communication links to work bidirectional, that is, each edge represents two *links*, one in each direction. The *bandwidth* of a link is defined as the number of *channels* it supports. A *packet* is defined as the amount of data that can be sent along a channel in one time unit. Hence the number of packets a link can forward in one time step is at most the number of channels it supports. Unless explicitly mentioned we assume that the bandwidth of the links is one. The *size* N of G is defined as the number of nodes G contains.

Each node in V contains an *injection buffer* and a *delivery buffer*. The task of the injection buffer of a node v is to store all messages v wants to send out, and the task of the delivery buffer is to store all messages that are destined for v. Further we assume that each link has a *link buffer* (or *buffer* for short). The *size* of a buffer is defined as the maximal number of packets a buffer can store. (Note that a message may consist of several packets.) In the *multi-port* model each node can forward packets simultaneously along all of its outgoing links in one time step, whereas in the *single-port* model each node can forward packets along at most one of its outgoing links in one time step. We only consider the multi-port model in the following, since it has become most common for packet routing in the last years.

The following two items are important for a network to have a good communication performance.

- The diameter should be small, since it is a lower bound for the worst-case time necessary to send messages between pairs of nodes that have maximal distance. If the degree is constant then the best diameter that can be reached for a graph of size N is $O(\log N)$.
- The network should have no "bottleneck" that forces many messages to traverse a small set of edges. The complete binary tree, for instance, is a very bad communication network since one half of its nodes can only communicate with the other half via its root.

In order to measure the second parameter, the bisection width or expansion of a network is often considered. For every $U, U' \subseteq V$, let $\bar{U} = V \setminus U$ and $C(U, U')$ be the number of edges in E that have one endpoint in U and one endpoint in U'. The *bisection width* of G is defined as

$$\min_{U \subseteq V, |U| = \lfloor |V|/2 \rfloor} \frac{C(U, \bar{U})}{|U|} .$$

A more general form is the *flux* [LR88] or *expansion* of a graph G. It is defined as

$$\min_{U \subseteq V, |U| \le \lfloor |V|/2 \rfloor} \frac{C(U, \bar{U})}{|U|} .$$

Network Topologies. The most commonly used networks in the routing literature and in parallel computers are meshes, tori and butterflies. These classes are defined in the following.

Definition 3.4.1 (Butterfly). *Let* $d \in \mathbb{N}$. *The* d-*dimensional butterfly* $BF(d)$ *is defined as an undirected graph with node set* $V = [d+1] \times [2]^d$ *and an edge set* $E = E_1 \cup E_2$ *with*

$$E_1 = \{\{(i,\alpha),(i+1,\alpha)\} \mid i \in [d], \ \alpha \in [2]^d\}$$

and

$$E_2 = \{\{(i,\alpha),(i+1,\beta)\} \mid i \in [d], \ \alpha,\beta \in [2]^d, \ \alpha \ and \ \beta \ differ$$
$$only \ at \ the \ ith \ position\} \ .$$

The d-*dimensional wrap-around butterfly* W-$BF(d)$ *is defined by taking the* $BF(d)$ *and identifying level* d *with level* 0.

Figure 3.1 shows the 3-dimensional butterfly $BF(3)$. $BF(d)$ has $(d+1)2^d$ nodes, $2d \cdot 2^d$ edges and degree 4.

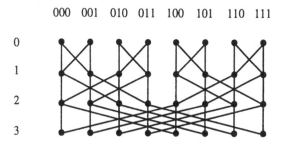

Fig. 3.1. The structure of BF(3).

The wrap-around butterfly is related to the following graph.

Definition 3.4.2 (Shuffle-Exchange). *Let* $d \in \mathbb{N}$. *The* d-*dimensional shuffle-exchange* $SE(d)$ *is defined as an undirected graph with node set* $V = [2]^d$ *and an edge set* $E = E_1 \cup E_2$ *with*

$$E_1 = \{\{(a_{d-1}\ldots a_0),(a_{d-1}\ldots \bar{a}_0)\} \mid (a_{d-1}\ldots a_0) \in [2]^d, \ \bar{a}_0 = 1 - a_0\}$$

and

$$E_2 = \{\{(a_{d-1}\ldots a_0),(a_0 a_{d-1}\ldots a_1)\} \mid (a_{d-1}\ldots a_0) \in [2]^d\} \ .$$

Figure 3.2 shows the 3- and 4-dimensional shuffle-exchange graph.

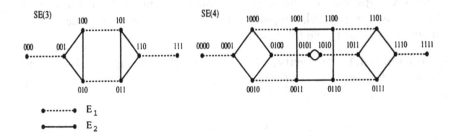

Fig. 3.2. The structure of SE(3) and SE(4).

Definition 3.4.3 (Torus, Mesh). *Let $m, d \in \mathbb{N}$. The (m, d)-mesh $M(m, d)$ is a graph with node set $V = [m]^d$ and edge set*

$$E = \left\{ \{(a_{d-1} \ldots a_0), (b_{d-1} \ldots b_0)\} \mid a_i, b_i \in [m], \sum_{i=0}^{d-1} |a_i - b_i| = 1 \right\} .$$

The (m, d)-torus $T(m, d)$ consists of an (m, d)-mesh and additionally an edge from $(a_{d-1} \ldots a_{i+1} 0\, a_{i-1} \ldots a_0)$ to $(a_{d-1} \ldots a_{i+1}(m - 1)a_{i-1} \ldots a_0)$ for every $i \in [d]$ and $a_j \in [m]$. $M(m, 1)$ is also called linear array, *$T(m, 1)$* cycle, *and $M(2, d) = T(2, d)$ d-dimensional hypercube.*

Figure 3.3 presents a linear array, a torus, and a hypercube.

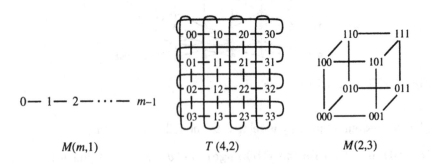

Fig. 3.3. The structure of $M(m, 1)$, $T(4, 2)$, and $M(2, 3)$.

Next we define the class of node-symmetric graphs.

Definition 3.4.4 (Node-Symmetric Graph). *A graph $G = (V, E)$ is called node-symmetric if for any pair of nodes $u, v \in V$ there exists an isomorphism $\varphi : V \to V$ with $\varphi(u) = v$ such that the graph $G_\varphi = (V, E_\varphi)$ with $E_\varphi = \{\{\varphi(x), \varphi(y)\} \mid \{x, y\} \in E\}$ is equal to G.*

Intuitively, node-symmetry means that a graph looks the same from any node. Node-symmetric graphs form a very general class and include most of the standard networks such as the d-dimensional torus, the wrap-around butterfly, the hypercube, etc. Another important class of graphs are the edge-symmetric graphs.

Definition 3.4.5 (Edge-Symmetric Graph). *A graph $G = (V, E)$ is called edge-symmetric if for any pair of edges $\{u, v\}, \{u', v'\} \in E$ there exists an isomorphism $\varphi : V \to V$ with $\{\varphi(u), \varphi(v)\} = \{u', v'\}$ such that the graph $G_\varphi = (V, E_\varphi)$ with $E_\varphi = \{\{\varphi(u), \varphi(v)\} \mid \{u, v\} \in E\}$ is equal to G.*

Note that every regular edge-symmetric graph is also node-symmetric. The d-dimensional torus, for instance, is edge-symmetric. Another important class of graphs are the leveled graphs.

Definition 3.4.6 (Leveled Graph). *A graph $G = (V, E)$ is called* leveled *with depth D if the nodes of G can be partitioned into $D+1$ levels $L_0, \ldots, L_D$ such that every edge in E connects nodes of consecutive levels. Nodes in level 0 are called* inputs, *and nodes in level D are called* outputs. *If, in addition, $|L_0| = |L_D|$ and L_0 is identified with L_D, then G is called a* wrapped leveled graph *with depth D.*

The d-dimensional butterfly, for instance, is a leveled graph with depth d, and the d-dimensional wrap-around butterfly is a wrapped leveled graph with depth d. Another important class of graphs are the expanders.

Definition 3.4.7 (Expander). *A graph is called* expander *if it has a constant expansion.*

Note that there exist expanders of constant degree. So far, the best expanders that have an explicit construction are all node-symmetric (see, e.g., [LPS88, Ma88, Mo94]).

3.4.2 The Routing Problem

There are two kinds of routing problems studied in the literature: static and dynamic routing. In case of static routing it is assumed that all messages are given in advance, whereas in case of dynamic routing it is assumed that new messages are continuously injected into the network. Hence within the dynamic routing model one is primarily interested in protocols that can sustain a certain arrival rate of messages, whereas within the static routing model one is primarily interested in protocols that can deliver the messages as fast as possible. In this book we only deal with static routing problems. This type of problems is defined as follows.

Consider an arbitrary network G with node set $[N]$. An instance of the static routing problem is defined by a multi-set

$$\mathcal{R} = \{(s_1, D_1), \ldots, (s_n, D_n)\} \ ,$$

where each pair $(s_i, D_i) \in [N] \times (\mathcal{P}([N]) \setminus \{\emptyset\})$ represents a message that has to be sent from node s_i to all nodes in D_i. The following routing problems are usually studied.

- *Permutation routing:* Let S_N be the set of all permutations $\pi : [N] \to [N]$. Given a permutation $\pi \in S_N$, route a message from node i to node $\pi(i)$ for all $i \in [N]$.
- *h-function routing:* Let $F_{h,N}$ be the set of all h-functions $f : [h] \times [N] \to [N]$. Given an h-function $f \in F_{h,N}$, route a message from node i to node $f(j,i)$ for all $i \in [N]$ and $j \in [h]$. A 1-function is simply called *function*.
- *h-relation routing:* An h-relation is defined as an h-function $f \in F_{h,N}$ with $|f^{-1}(i)| = h$ for all $i \in [N]$. Given an h-relation $f \in F_{h,N}$, route a message from node i to node $f(j,i)$ for all $i \in [N]$ and $j \in [h]$.
- *Broadcasting:* Given a node $i \in [N]$, route the same message from node i to all nodes $j \in [N]$.
- *Gossiping:* For every node $i \in [N]$, route the same message from node i to all nodes $j \in [N]$.
- *Total exchange:* For every node $i \in [N]$, route a distinct message from node i to node j for all $j \in [N]$.

The solution to a routing problem is called *routing protocol*. It determines for each time step which message to forward along which link (or set of links). Its basic component is the *contention resolution rule* that decides which messages win if too many messages try to use the same link at the same time.

3.4.3 Message Passing Models

In the following we consider a message to have the following format.

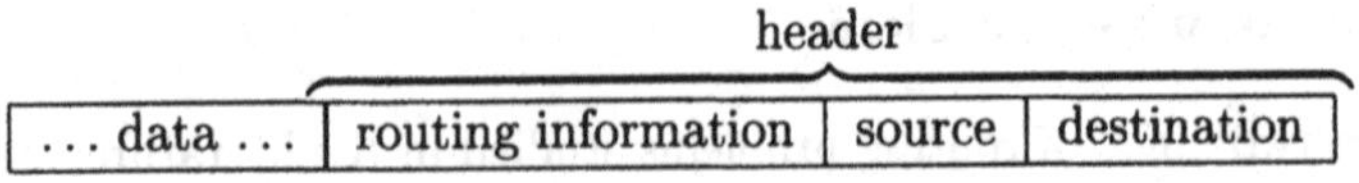

	header		
... data ...	routing information	source	destination

Fig. 3.4. The format of a message.

In a network with N nodes the source and destination need $\lceil \log N \rceil$ bits, each, to have a unique identification number. Throughout this paper we restrict the routing information to be very small, namely of length at most $O(\log N)$. It is usually needed to store information about the path a message has to follow or the priority level of a message. It may also be used for error detection. The routing information, source, and destination together form the *header* of a message. In contrast to the header of a message, its data field might be of arbitrary size. Unless explicitly mentioned, we assume in the following that the data fields of all messages that have to be routed according to some routing problem $\mathcal{R}$ are of the same size. In this book, we study the following two models: the *store-and-forward routing* model and the *wormhole routing* model.

Store-and-Forward Routing. The most commonly used and well understood routing model in the literature is the store-and-forward routing model. In this model time is partitioned into synchronous *steps*. One step is defined as the time a message needs to be sent along a link. (It is usually assumed that every link needs the same amount of time to forward a message. Furthermore, internal computations of processors are usually not counted as time steps as long as they are bounded by some small polynomial, since it is often assumed that the time to send a message along a link is much higher than the time needed to execute internal instructions.) A node must store an entire message before it can forward any part of the message along the next link on its route. Hence messages can be viewed as single packets.

Wormhole Routing. In wormhole routing, messages are sent as *worms*, each of which consists of a sequence of fixed size packets called *flits*. A flit is defined as the amount of data that can be sent along a channel in one time step. The *length* of a worm is the number of flits it contains. We call the first flit of a worm *head* and the remaining flits *body*. The first flits of a worm store the header and the remaining flits the data field of a message. In order to keep the routing hardware in a node small, the capacity of each link to buffer flits is usually limited to at most one per outgoing channel. Because a link can not identify which worm a flit belongs to from its contents, the flits of a worm must arrive on the incoming channel in a contiguous fashion. Since each channel can only buffer one flit of a worm, if the head of a worm can not advance then the flits following the head must stall.

Note that *circuit-switching* (i.e., establishing a connection between pairs of nodes in a network) can be viewed as a special case of wormhole routing in which the length of each worm is at least as large as the distance between the worm's source and its destination.

3.4.4 Routing Strategies

We distinguish between two kinds of protocols: *Offline* protocols and *online* protocols. In case of offline protocols, we allow all processors of a parallel system to have complete knowledge about the routing problem. This enables them in principle to develop an optimal strategy for sending the messages to their destinations (note that we do not count internal computations, but only communication rounds). In this area, important questions are, how fast such a strategy can be, and whether (asymptotically) optimal strategies can be computed in polynomial time. Note that offline protocols are also referred to as *global control* or *centralized protocols*.

In case of online protocols, at the beginning every processor only knows the messages that are currently stored in it (and, maybe, a few global parameters such as the size of the routing problem). In general, it makes no sense to inform all processors about the whole routing problem before starting to deliver the messages, since – as we will see – for many situations knowledge about a small fraction of the routing problem already suffices for the processors to develop very efficient routing strategies. Note that online protocols are also sometimes called *local control* or *distributed protocols*.

There are basically two classes of online protocols. In order to send a message from its source to its destination in G it has to traverse a contiguous sequence of links called *routing path*. Given a source-destination pair a message may either have to traverse a path specified in advance or is able to choose among several alternative paths depending on the source-destination pairs of other messages or other events. The first case is called *oblivious routing* and the second case *adaptive routing*.

Oblivious Routing. In case of oblivious routing the path a message uses only depends on its source and destination. Let $p_{v,w}$ denote the path from v to w in G that has to be taken by every message that wants to travel from v to w. Then $\mathcal{P} = \{p_{v,w} \mid v, w \in V\}$ is called *path system* of size $|V|$. In this case any routing problem $\mathcal{R}$ can be defined by specifying a *path collection* $\mathcal{P}_\mathcal{R}$ of size $|\mathcal{R}|$ that contains the path $p_{v,w}$ for every pair (v, w) in $\mathcal{R}$. $\mathcal{P}_\mathcal{R}$ is called

– *simple* if no path in $\mathcal{P}_\mathcal{R}$ contains the same edge more than once,
– *shortcut-free* if no piece of a path in $\mathcal{P}_\mathcal{R}$ can be shortcut by any combination of other pieces of paths in $\mathcal{P}_\mathcal{R}$,
– *shortest* if all paths in $\mathcal{P}_\mathcal{R}$ are shortest paths in G, and
– *leveled* if the nodes can be arranged in levels such that for every pair (v, w) in $\mathcal{R}$, the path from v to w in $\mathcal{P}_\mathcal{R}$ only contains edges that lead from level i to level $i + 1$ for some $i \geq 0$.

The following relationship holds for these types of path collections.

$$\text{leveled p. c.} \subset \text{shortest p. c.} \subset \text{shortcut-free p. c.} \subset \text{simple p. c.}$$

Important parameters for the runtime are

– the *size n* of $\mathcal{P}_\mathcal{R}$, that is, the number of paths $\mathcal{P}_\mathcal{R}$ contains,
– the *congestion C* of $\mathcal{P}_\mathcal{R}$, that is, the maximal number of paths in $\mathcal{P}_\mathcal{R}$ that contain the same edge in G, and
– the *dilation D* of $\mathcal{P}_\mathcal{R}$, that is, the length of a longest path in $\mathcal{P}_\mathcal{R}$.

In case that we want to route random functions, we often use the notion of expected congestion that is defined as follows. For an edge e and a path system $\mathcal{P}$, let the expected congestion $\bar{C}_e$ of e be defined as the expected number of messages that traverse e using paths in $\mathcal{P}$ during the routing of a randomly chosen $(1-)$function. The *expected congestion* of $\mathcal{P}$ is then defined as $\bar{C} = \max_{e \in E} \bar{C}_e$.

Adaptive Routing. In case of adaptive routing, the path a message uses is not predetermined. There are several parameters that are used to measure the performance of adaptive routing protocols: The bisection width, flux, expansion, or routing number (see Chapter 5) of a network.

In adaptive protocols that do not restrict the messages to approach their destinations via shortest paths, *livelocks* can happen. Messages are defined to run into a livelock if they run infinitely often along the same cycle in the network.

3.4.5 Space-Efficient Routing

In theory it is often assumed that we have unit size messages, unbounded buffers, and enough space in the processors to store the protocol and routing information such as routing tables that assign to each source-destination pair a subset of links leaving that processor. In practice, however, messages are only allowed to have a header of limited size (ATM-packets, e.g., only consist of 53 bytes [Bl95]), buffers can only store a limited number of messages (or even a limited number of parts of a message), and processors only have a limited amount of space to store routing information. Within the field of space-efficient routing there are two mainstreams.

Routing with Bounded Buffer Size

In networks with bounded buffers it may happen that messages prevent each other from moving forward. Such a situation is called *deadlock*. It is a challenging task to find oblivious and adaptive protocols that run efficient and can avoid deadlocks in case of bounded buffers. Protocols that do not require any edge buffers to route the messages to their destinations are called *hot potato* routing protocols.

Compact Routing

The theory of *compact routing* deals with questions about how a limited space at the processors influences the efficiency of routing tables. It basically follows two directions.

Relationship between Space and Stretch Factor. The efficiency of a routing scheme is measured in terms of its *stretch factor*, that is, the maximum ratio over all pairs of nodes between the length of a route produced by the scheme and the length of a shortest path between these nodes. Let G be an arbitrary network, and let v be any node in G. Given any $s \geq 1$, let $\mathcal{S}_s(G)$ denote the class of all path selection strategies with stretch factor s in G. For every strategy $S \in \mathcal{S}_s(G)$, let $MEM(v, S)$ be the *memory requirement* in v to realize strategy S. Then we define, for every network $G = (V, E)$ and strategy S,

- the *global memory requirement* of S in G as
  $MEM_g(G, S) = \sum_{v \in V} MEM(v, S)$, and
- the *local memory requirement* of S in G as
  $MEM_\ell(G, S) = \max_{v \in V} MEM(v, S)$.

Using this, we define, for every stretch factor s,

- the *global memory requirement* of G as
  $MEM_g(G, s) = \min_{S \in \mathcal{S}_s(G)} MEM_g(G, S)$, and
- the *local memory requirement* of G as
  $MEM_\ell(G, s) = \min_{S \in \mathcal{S}_s(G)} MEM_\ell(G, S)$.

It is interesting to consider both global and local memory requirements because it is possible that $MEM_g(G, s) \ll nMEM_\ell(G, s)$ for some graph G and stretch factor s. That is, informally, the global memory requirement does not indicate whether the routing information can be evenly balanced between all nodes. On the contrary, the local memory requirement does not indicate whether all nodes need such a large memory.

The space for storing information in the headers of messages is usually bounded to be at most logarithmic to the size of G.

Relationship between Space and Slowdown. Within this model the efficiency of a routing scheme is measured in terms of its *slowdown*, that is, the ratio between the worst case time to route any permutation in a network using the scheme and the routing number of the network (see Section 5 for a definition of the routing number). Given a slowdown s, the local and global space requirements $MEM_g(G, s)$ and $MEM_\ell(G, s)$ of a scheme are measured as defined for the stretch factor above.

4. Introduction to Store-and-Forward Routing

In the following six chapters we will concentrate on presenting store-and-forward routing strategies. Such strategies are used on machines such as the NCube, NASA MPP, Intel Hypercube, and Transputer-based machines. Since the store-and-forward model assumes that all messages are of unit length, it is the easiest and therefore the most studied model in the literature.

Let us start with giving an overview of results in the area of store-and-forward routing, and describing networks for which optimal randomized and deterministic store-and-forward routing protocols are already known. Chapter 5 introduces a parameter called the routing number. This parameter will be used extensively in the following chapters to apply routing protocols to arbitrary networks. In Chapter 6 we prove the existence of three offline protocols for routing packets along a fixed path collection and show, how these protocols can be applied to network emulations. Chapter 7 gives an overview on the best universal oblivious protocols for store-and-forward routing known so far. It is shown, how each of these protocols can be applied to routing in specific networks, and what the limitations of these protocols are. Chapter 8 gives an overview on adaptive store-and-forward routing protocols, and describes several techniques for developing universal adaptive protocols. It is further demonstrated how efficiently these techniques can be applied to routing in arbitrary networks. In Chapter 9 we show how a limited space for storing routing information in the packets and the nodes of the network influences the performance of routing algorithms. Both randomized and deterministic protocols are developed that can be efficiently applied to arbitrary networks even under severe space limitations.

4.1 History of Store-and-Forward Routing

In this section we give an overview on results known about store-and-forward routing in specific networks and universal oblivious store-and-forward routing protocols.

4.1.1 Routing in Specific Networks

In 1965, Beneš [Be65] showed that the inputs and outputs of an N-node Beneš network (two back-to-back butterfly networks) can be connected in any permutation by a set of disjoint paths. Shortly afterwards, Waksman [Wa68] devised a simple sequential algorithm for finding the paths in $O(N)$ time. Given the paths, it is straightforward to route a set of packets from the inputs to the outputs of an N-node Beneš network in any one-to-one fashion in $O(\log N)$ steps using buffers of size 1. Although the inputs comprise only $O(N/\log N)$ nodes in an N-node Beneš network, it is possible to route $\log N$ permutations from the inputs to the outputs in $O(\log N)$ steps by routing the permutations one after the other in a continuous fashion (i.e., using *pipelining*). Unfortunately, no efficient parallel algorithm for finding the paths is known.

In 1968, Batcher [Ba68] devised an elegant and practical parallel algorithm for sorting N packets on an N-node shuffle-exchange network in $\log^2 N$ steps using buffers of size 1. The algorithm can be used to route any permutation of packets by sorting them according to their destination addresses. The result extends to routing many-one problems provided that (as is typically assumed) two packets with the same destination can be *combined* to form a single packet should they meet en route to their destination.

No better deterministic algorithm was found until 1983, when Ajtai *et al.* [AKS83] solved a classical open problem by constructing an $O(\log N)$-depth sorting network. Leighton [Le85] then used this $O(N \log N)$-node network to construct a degree 3 N-node network capable of solving any N-packet routing problem in $O(\log N)$ steps using buffers of size 1. Although this result is optimal up to constant factors, the constant factors are quite large and the algorithm is of no practical use. Hence, the effort to find fast deterministic algorithms continued. In 1989, Upfal discovered an $O(\log N)$-step algorithm for routing on an expander-based network called *multibutterfly* [Up89]. The algorithm solves the routing problem directly without reducing it to sorting, and the constant factors are much smaller than those of the AKS-based algorithms. In [LM92], Leighton and Maggs show that the multibutterfly is fault tolerant and improve the constant factors in Upfal's analysis. Borodin *et al.* [BRSU93] further show that any permutation can be routed deterministically in an s-ary multibutterfly of size N in time $O(\log_s N)$. Recently, Maggs and Vöcking [MV97] found a constant degree variant of the multibutterfly that allows every h-relation to be routed in $O(h + \log N)$ steps.

There has also been great success in the development of efficient randomized packet routing algorithms. The study of randomized algorithms was pioneered by Valiant [Va82] who showed how to route any permutation of N packets in $O(\log N)$ steps on an N-node hypercube with buffers of size $O(\log N)$ at each node. Valiant's idea was to route each packet to a randomly chosen intermediate destination before routing it to its true destination. Although the algorithm is not guaranteed to deliver all packets within $O(\log N)$

steps, for any permutation it does so with high probability. In particular, the probability that the algorithm fails to deliver the packets within $O(\log N)$ steps is at most $1/N^k$, for any constant $k > 0$. (The value of k can be made arbitrarily large by increasing the constant in the $O(\log N)$ bound.) In the following, we use the term "w.h.p." (with high probability) whenever we have a probability of at least $1 - 1/N^k$ for any constant $k > 0$.

Valiant's result was improved in a succession of papers by Aleliunas [Al82], Upfal [Up82], Pippenger [Pi84], and Ranade [Ra87]. Aleliunas and Upfal developed the notion of a *delay path* and showed how to route on the shuffle-exchange and butterfly of size N in $O(\log N)$ steps, w.h.p., using buffers of size $O(\log N)$. Pippenger was the first to eliminate the need for large buffers, and showed how to route on a variant of the butterfly in $O(\log N)$ steps, w.h.p., with buffers of size $O(1)$ [Pi84]. Ranade showed how combining can be used to extend the Pippenger result to include many-one routing problems, and tremendously simplified the analysis required to prove such a result. As a consequence, it has finally become possible to simulate a step of an N-processor CRCW PRAM on an N-node butterfly or hypercube in $O(\log N)$ steps, w.h.p., using constant-size buffers on each edge [Ra87]. Borodin *et al.* [BRSU93] further showed that any permutation can be routed in any s-ary butterfly of size N in time $O(\log_s N)$, w.h.p.

Concurrent with the development of these hypercube-related packet routing algorithms has been the development of algorithms for routing in meshes. The randomized algorithm of Valiant and Brebner can be used to route any permutation of N packets on a $(\sqrt{N}, 2)$-mesh in $O(\sqrt{N})$ steps using buffers of size $O(\log N)$. Kunde [Ku88] showed how to route deterministically in $(2 + \epsilon)\sqrt{N}$ steps using buffers of size $O(1/\epsilon)$. Krizanc *et al.* [KRT88] showed how to route any permutation in $2\sqrt{N} + O(\log N)$ steps, w.h.p., using constant-size buffers. Furthermore, Leighton *et al.* [LMT89] discovered a deterministic algorithm for routing any permutation in $2\sqrt{N} - 2$ steps using constant-size buffers, thus achieving the optimal time bound in the worst case. In case of d-dimensional meshes, the results by Kunde [Ku91] and Suel [Su94] imply that there is a deterministic protocol for routing any permutation in optimal time using constant size buffers. Kaklamanis *et al.* [KKR93a] present a randomized hot-potato protocol that routes any permutation and randomly chosen function in a d-dimensional torus of side length n in time $d \cdot n + O(\log^2 n)$, w.h.p.

4.1.2 Universal Routing

In the last years, also *universal* routing protocols have been developed, that is, protocols that can be used for any communication pattern in any network. The advantage of these protocols is that they can be quickly adapted to topology changes that occur if new processors are added or some break down. In 1988, Leighton *et al.* [LMR88] showed that for any simple path collection with congestion C and dilation D there is an offline routing scheme that routes a packet along each of the paths in time $O(C + D)$ using (sufficiently large) constant size edge buffers. In case that only buffers of size 1 are allowed, Meyer auf der Heide and Scheideler [MS95b] showed that packets can be routed offline along an arbitrary simple path collection with congestion C and dilation D in time $O((D+C\log(C+D))\log\log(C+D))(\log\log\log(C+D))^{1+\epsilon})$ for any constant $\epsilon > 0$.

Several results about online universal protocols have also been found. Leighton *et al.* [LMR88] presented an online protocol for routing packets in any simple path collection of size n with congestion C and dilation D in time $O(C + D\log(nD))$, w.h.p. In another paper, Leighton *et al.* [LMRR94] used the techniques in [Ra87] to develop an online protocol that can route packets along any leveled path collection of size n with depth D and congestion C in time $O(C + D + \log n)$, w.h.p, using constant size buffers. Meyer auf der Heide and Vöcking [MV95] presented a protocol that can route packets along any shortcut-free path collection of size n with congestion C and dilation D in time $O(C + D + \log n)$, w.h.p., using buffers of size C. Cypher *et al.* [CMSV96] developed two randomized protocols for arbitrary simple path collections. The first protocol runs in time $O((D\log\log n + C + \frac{\log n \cdot \log\log n}{\log\log(C\cdot D)}) \cdot \frac{\log(C\cdot D)}{\log\log(C\cdot D)})$, w.h.p., and requires buffers of size $\Theta(\frac{\log(C\cdot D)}{\log\log(C\cdot D)})$. The second protocol runs in time $O(C \cdot D^{1/B} + D\log n)$, w.h.p., if buffers of size B are available. Furthermore, Rabani and Tardos [RT96] showed that there is a randomized protocol for arbitrary simple path collections that runs in time $O(C) + (\log^* n)^{O(\log^* n)} D + poly(\log n)$, w.h.p., using buffers of size $poly(\log n)$. Recently, Ostrovsky and Rabani [OR97] presented a randomized protocol for arbitrary simple path collections that runs in time $O(C + D + \log^{1+\epsilon} n)$ for any constant $\epsilon > 0$, w.h.p., using buffers of size $poly(\log n)$.

Meyer auf der Heide and Scheideler [MS96a] developed a universal deterministic routing protocol. They showed, for instance, that for any network with sufficiently large diameter (such as the class of planar networks), this protocol reaches the best possible worst case time for routing arbitrary permutations. A detailed description of the universal protocols mentioned above will be given in Chapters 6 to 8.

4.2 Optimal Networks for Permutation Routing

Consider an arbitrary network of size N with degree d. Clearly, since the diameter of any graph of size N with degree d is at least $\log_d N$, there exists a permutation that takes at least $\log_d N$ steps to be routed. The question is whether there are networks that can reach this lower bound up to constants. In the following we describe two classes of networks that either have an optimal randomized routing protocol or an optimal deterministic routing protocol.

4.2.1 Optimal Networks for Randomized Routing

In this section we show that for any $d \geq 4$ there exists a network of size N with degree d such that there is a randomized protocol for routing an arbitrary permutation in optimal time $O(\log_d N)$, w.h.p. For this we introduce the s-ary butterfly.

Definition 4.2.1 (s-ary Butterfly). *Let $s, d \in \mathbb{N}$. The s-ary d-dimensional butterfly (s, d)-BF is defined as an undirected graph with node set $V = [d+1] \times [s]^d$ and an edge set*

$$ E = \bigcup_{i=0}^{s-1} \{\{(\ell, x), (\ell+1, f(x, \ell, i))\} \mid \ell \in [d],\ x \in [s]^d\}, $$

where $f(x, \ell, i)$ is defined as $f(x, \ell, i) = (x_{d-1}, \ldots, x_{\ell+1}, i, x_{\ell-1}, \ldots, x_0)$.

The s-ary d-dimensional wrap-around butterfly (s, d)-WBF is defined by taking the (s, d)-BF and identifying level d with level 0.

Note that the $(n, 1)$-WBF is the complete graph consisting of n vertices. The following picture gives an example of an (s, d)-BF.

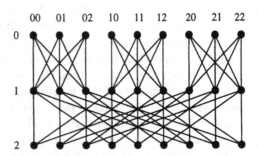

Fig. 4.1. The structure of a $(3, 2)$-BF.

The following theorem can be shown. Its proof can be found in [BRSU93].

Theorem 4.2.1. *For any $d \geq 1$ and $s \geq 2$, there exists a protocol that can route any permutation in the s-ary d-dimensional butterfly of size N in time $O(\log_s N)$, w.h.p.*

4.2.2 Optimal Networks for Deterministic Routing

In this section we show that for sufficiently large d there exists a network of size N with degree d such that there is a deterministic protocol for routing an arbitrary permutation in optimal time $O(\log_d N)$. For this we introduce the (elementary) s-ary multibutterfly.

The basic building block of the s-ary multibutterfly is an s-ary m-splitter (see [BRSU93]).

The s-ary m-splitter (or (s, m)-splitter) is a bipartite graph with m input nodes and m output nodes. In this graph the output nodes are partitioned into $\sqrt{s}$ output sets, each with $m/\sqrt{s}$ nodes. Every input node has $\frac{\sqrt{s}}{2}$ edges to each of the $\sqrt{s}$ output sets. The edges connecting the input set to each of the output sets define an expander graph with properties described in [BRSU93].

The s-ary d-dimensional multibutterfly (s, d)-MBF has $d + 1$ levels. The vertices at level $0 \leq i \leq d - 1$ are partitioned into $\sqrt{s}^i$ sets of $m_i = \sqrt{s}^{d-i}$ consecutive nodes. Each of these sets in level i is an input set of an s-ary m_i-splitter. The output sets of that splitter are $\sqrt{s}$ sets of size m_{i+1} in level $i + 1$. Thus each node in the (s, d)-MBF is the endpoint of at most $2 \cdot \frac{s}{2} = s$ edges. Figure 4.2 shows the structure of a $(16, d)$-MBF.

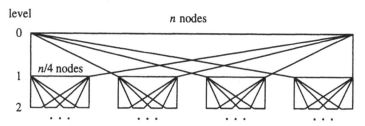

Fig. 4.2. The structure of a $(16, d)$-MBF.

Using the (s, d)-MBF, the following result can be shown. Its proof can be found in [BRSU93].

Theorem 4.2.2. *For sufficiently large s, the s-ary multibutterfly of size N can route any permutation from the inputs to the outputs in time $O(\log_s N)$.*

Using techniques in [Up92], the s-ary multibutterfly can be extended to route any global permutation in $O(\log_s N)$ steps.

5. The Routing Number

In the following chapters we will often refer the routing time needed by our online routing protocols to the worst case routing time of a best offline routing algorithm, the so-called *routing number*. This number is defined as follows (see, e.g., [ACG94, MS96a]):

Consider an arbitrary network G with N nodes and link bandwidth one. For a permutation $\pi \in S_N$, let $R(G, \pi)$ be the minimum possible number of steps required to route packets offline in G according to π using the multi-port model with unbounded buffers. Then the *routing number* $R(G)$ of G is defined by

$$R(G) = \max_{\pi \in S_N} R(G, \pi) \ .$$

In case that there is no risk of confusion about the network G we will write R instead of $R(G)$. The routing number has the following nice property.

Theorem 5.0.3. *For any network G with routing number R and any routing strategy, the average number of steps to route a permutation in G is bounded by $\Omega(R)$.*

Proof. Let $\bar{R} = \frac{1}{|S_N|} \sum_{\pi \in S_N} R(G, \pi)$ denote the average number of steps to route a permutation in G using the best possible routing strategy. Consider any fixed permutation π. In order to bound the minimum number of steps to route π we will use a probabilistic argument based on Valiant's trick (see [Va82]) by first sending the packets to random intermediate destinations before sending them to the destinations prescribed by π. Let X be a random variable denoting the minimum number of steps necessary to route a randomly chosen permutation. According to the Markov Inequality it holds:

$$\Pr[X \geq 3\bar{R}] \leq \tfrac{1}{3} \ .$$

Therefore, for a randomly chosen permutation φ for the intermediate destinations, it holds that the minimum number of steps to route the packets first to their intermediate destinations (prescribed by φ) and then to their final destinations (prescribed by π) is at least $6\bar{R}$ with probability at most $\tfrac{1}{3} + \tfrac{1}{3} < 1$. Therefore there exists an offline protocol that routes π in at most $6\bar{R}$ steps. Thus $R \leq 6\bar{R}$, which completes the proof. $\qquad\square$

Hence asymptotically the routing number is not only an upper bound, but also a lower bound for the average permutation routing time using optimal routing strategies in G. This demonstrates that the routing number is a robust measure for the routing performance of networks.

Note that the routing number might be defined by some protocol which uses specific routing paths tailored to the permutation to be routed. This implies the following result.

Remark 5.0.1. For any network G of size N with routing number R, there exists a collection of simple paths for any permutation routing problem $\pi \in S_N$ with congestion at most R and dilation at most R.

The question is whether it is possible to find such a path collection in an efficient way for every permutation routing problem. This will be answered in the following two sections. In the first section we prove the existence of path systems with low dilation and expected congestion, and in the second section we show how to choose path collections out of such systems in a distributed way such that the dilation and congestion bounds in Remark 5.0.1 can be reached up to constant factors for any permutation routing problem. Furthermore we show in Section 5.4 how the routing number of a network is related to its expansion. Section 5.3 gives bounds on the routing number of some important classes of networks. Finally, Section 5.5 contains high-level descriptions of algorithms for constructing efficient path systems in arbitrary networks.

5.1 Existence of Efficient Path Systems

A necessary prerequisite for efficient oblivious routing is a path system with low dilation and low expected congestion. In this section we construct path systems whose dilation and expected congestion are close to the routing number for arbitrary networks. The main result of this section is formulated in the next theorem.

Theorem 5.1.1. *For any network G of size N with routing number R there is a simple path system with dilation at most R and expected congestion at most R. Furthermore, for random functions the congestion is bounded by $R + O(\sqrt{\max\{R, \log N\} \cdot \log N})$, w.h.p.*

Proof. Let G be a network of size N with routing number R. Then for any permutation $\pi_i : [N] \to [N]$ with $\pi_i(x) = (x + i) \bmod N$ for all $i, x \in [N]$ there is a path collection $\mathcal{P}_i$ along which packets can be routed in at most R time steps. Therefore the congestion and dilation of $\mathcal{P}_i$ is at most R. We then choose $\mathcal{P} = \bigcup_{i=0}^{N-1} \mathcal{P}_i$ to be the path system for G. Clearly, this path system has congestion at most $N \cdot R$ and dilation at most R. It remains to bound the expected congestion and the congestion that holds w.h.p. for routing a random function in G using paths in $\mathcal{P}$.

Consider any edge e in G. Let the random variable X denote the number of paths that cross e and are used by a packet. Since each node in G chooses a destination for its packet uniformly and independently at random, the probability that a path crossing e is used by a packet is $1/N$. Therefore the expected congestion is at most R.

For any node v in G, let the binary random variable X_v be one if and only if the packet with source v contains e in its routing path. Then $X = \sum_{v \in V} X_v$. Since we consider routing a random function, the probabilities of all X_v are independent. As $E[X] \leq R$, the Chernoff bounds (see Theorem 3.2.2) therefore yield that

$$\Pr[X \geq (1+\epsilon)R] \leq \begin{cases} e^{-\epsilon^2 R/2} & : \quad \text{if } 0 \leq \epsilon \leq 1 \\ e^{-\epsilon \cdot R/3} & : \quad \text{if } \epsilon > 1 \end{cases}$$

This probability is polynomially small in N for $\epsilon = O(\max\{\sqrt{\frac{\log N}{R}}, \frac{\log N}{R}\})$ sufficiently large. Hence w.h.p. all edges have a congestion of at most $R + O(\sqrt{\max\{R, \log N\} \cdot \log N})$. □

For node-symmetric networks we can strengthen Theorem 5.1.1 by showing that the path system may even consist solely of shortest paths. This is important, because some oblivious routing protocols are especially good for shortest path systems, see Chapter 7. Since the diameter of a network is a lower bound for its routing number, we use the diameter instead of the routing number in the following theorem.

Theorem 5.1.2. *Let G be a node-symmetric network with N nodes and diameter D. Then there exists a shortest path system that has an expected congestion of $D + O(\sqrt{\frac{D \log N}{N}})$.*

Proof. Consider the problem of gossiping in G. For every source-destination pair $(u, w) \in V^2$, perform the random experiment of choosing uniformly at random one path out of the set of all shortest paths that connect u with w. For every node v in G, let the binary random variable $X_{u,w}^v$ be one if and only if the path chosen from u to w traverses v, and $p_{u,w}^v = \Pr[X_{u,w}^v = 1]$. Further let the random variable C_v denote the number of paths traversing v. Then it holds

$$C_v = \sum_{(u,w) \in V^2} X_{u,w}^v .$$

Since G is node-symmetric, there exists for every node v' in G an automorphism φ that maps v to v'. The uniform choice of the paths ensures that $p_{u,w}^v = p_{\varphi(u),\varphi(w)}^{\varphi(v)}$ for all node pairs (u, w). Thus it holds

$$E[C_v] = \sum_{(u,w)\in V^2} p_{u,w}^v = \sum_{(u,w)\in V^2} p_{\varphi(u),\varphi(w)}^{\varphi(v)}$$

$$= \sum_{(u,w)\in V^2} p_{u,w}^{v'} = E[C_{v'}]$$

Hence, $E[C_v]$ is the same for every node v in G, namely at most $N \cdot D$. Since the paths are chosen independently at random, applying Chernoff bounds yields that $C_v = N \cdot D + O(\sqrt{D \cdot N \log N})$, w.h.p. Thus there exists a path system in G with such a congestion. Since, for a randomly chosen function, each of the paths has a probability of $\frac{1}{N}$ to be chosen, Theorem 5.1.2 follows.

$\square$

Theorem 5.1.2 implies, for instance, that for the wrap-around butterfly network there exists a shortest path system connecting all of its nodes with expected congestion at most its diameter plus one. It also implies how to construct such a path system: Just choose, for any pair of nodes, a path uniformly at random from all shortest paths connecting these nodes. Similarly, the following result holds for edge-symmetric networks.

Theorem 5.1.3. *Let G be a edge-symmetric network with N nodes, degree d, and diameter D. Then there exists a shortest path system that has expected congestion*

$$\tfrac{D}{d} + O(\tfrac{1}{N}\sqrt{\max\{\tfrac{N \cdot D}{d}, \log N\} \cdot \log N}) \ .$$

5.2 Valiant's Trick

In Section 5.1 we have seen that for any network with routing number R there is a fixed path system that yields asymptotically optimal parameters C and D for almost all functions. However, Borodin and Hopcroft [BH85] could prove a very high lower bound for the maximal congestion that can be reached by a permutation routing problem in a network using a fixed path system. Their result has been improved by Kaklamanis *et al.* [KKT91]. Together with an extension by Parberry [Pa90] we obtain the following theorem.

Theorem 5.2.1. *Let G be an arbitrary network of size N with degree d, and let $\mathcal{P}$ be an arbitrary path system in G. Let n nodes in G be determined as sources and destinations. Then there is a permutation $\pi \in S_n$ that has a congestion C_π of $\Omega(\frac{n}{d\sqrt{N}})$.*

Hence, if G has constant degree and all nodes in G are source and destination, that is $n = N$, then there exists a permutation with congestion $\Omega(\sqrt{N})$. Therefore the congestion might be much higher than the dilation, since networks of bounded degree can have a diameter of $O(\log N)$.

For the d-dimensional hypercube $M(2, d)$ Theorem 5.2.1 implies a worst case congestion of $C = \Omega(\sqrt{2^d}/d)$. Kaklamanis *et al.* present a path system in [KKT91] that reaches this bound.

In case that packets have to be sent from the inputs to the outputs of a d-dimensional wrap-around butterfly according to some arbitrary permutation, Bock [Bo96] presents a path system that reaches the lower bound $C = \Omega(\sqrt{2^d}/d)$.

According to Theorem 5.2.1, oblivious routing might perform very poorly for some functions. Thus, in order to get close to the routing number, we have to turn to non-oblivious strategies. A beautiful, simple idea was presented in [Va82].

> **Valiant's trick:**
> Consider routing an arbitrary permutation. Route the packets first to intermediate destinations, chosen uniformly and independently at random, before routing them to their true destinations.

Applying this trick to a fixed path system $\mathcal{P}$ yields the following strategy:

Each node first chooses random intermediate destinations for its packets. Then each packet is first sent to this intermediate destination, and from there to its final destination, both times using the path prescribed in $\mathcal{P}$.

With this strategy, the congestion of *every* permutation routing problem can be brought close to the expected congestion of $\mathcal{P}$, w.h.p. In particular, the following theorem holds.

Theorem 5.2.2. *Let $\mathcal{P}$ be an arbitrary path system of size n with dilation D and expected congestion $\bar{C}$. Then every permutation can be routed along paths in $\mathcal{P}$ using Valiant's trick with dilation at most $2D$ and congestion at most $2\bar{C} + O(\sqrt{\max\{\bar{C}, \log n\} \cdot \log n})$, w.h.p.*

Proof. Consider first the problem of routing a random function. Let V be the set of sources in $\mathcal{P}$. Consider an arbitrary edge e in $\mathcal{P}$. For every $v \in V$, let the binary random variable X_v be one if and only if the packet starting from v traverses e. Further let the random variable X denote the number of packets that traverse e. Then $X = \sum_{v \in V} X_v$. According to the definition of the expected congestion, it holds $E[X] \leq \bar{C}$. For a random function, the probabilities for all X_v are independent. Hence the Chernoff bounds yield that

$$\Pr[X \geq (1 + \epsilon)\bar{C}] \leq \begin{cases} e^{-\epsilon^2 \bar{C}/2} & : \text{ if } 0 \leq \epsilon \leq 1 \\ e^{-\epsilon \bar{C}/3} & : \text{ if } \epsilon > 1 \end{cases}$$

which is polynomially small in n for $\epsilon = O(\max\{\sqrt{\frac{\log n}{\bar{C}}}, \frac{\log n}{\bar{C}}\})$ sufficiently large. If we use Valiant's trick to route an arbitrary permutation then both situations, routing packets from their sources to intermediate destinations and from the intermediate destinations to their destinations, can be regarded as routing random functions. Hence the overall congestion of routing packets according to an arbitrary permutation using Valiant's trick is at most

$$2\bar{C} + O\left(\sqrt{\max\{\bar{C}, \log n\} \cdot \log n}\right) \; ,$$

w.h.p. Furthermore, the dilation of the resulting paths is at most $2D$.   □

An easy consequence of this theorem is the following corollary.

Corollary 5.2.1. *Let $\mathcal{P}$ be an arbitrary path system of size n with dilation D and expected congestion $\bar{C}$. Then every h-relation can be routed along paths in $\mathcal{P}$ using Valiant's trick with dilation at most $2D$ and congestion at most $2h \cdot \bar{C} + O(\sqrt{\max\{h\bar{C}, \log n\} \cdot \log n})$, w.h.p.*

Theorem 5.1.1 and Corollary 5.2.1 yield the following result.

Corollary 5.2.2. *For any network G of size N with routing number R there exists a simple path system $\mathcal{P}$, such that routing any h-relation along paths in $\mathcal{P}$ using Valiant's trick has dilation at most $2R$ and congestion at most $2h \cdot R + O(\sqrt{\max\{h \cdot R, \log n\} \cdot \log n})$, w.h.p.*

This result implies that if there is an oblivious protocol that can route packets along an arbitrary simple path collection with dilation D and congestion C in time $O(C + D)$ then a combination of the path selection strategy in Corollary 5.2.2 and this protocol would yield a routing protocol that reaches the optimal routing performance of permutation routing in arbitrary networks. We will show in Chapter 7 that in fact there is an oblivious routing protocol that comes quite close to this bound.

5.3 The Routing Number of Specific Networks

In this section we determine the routing number of some important classes of networks. Theorem 5.1.2 together with Theorem 5.2.2 and Theorem 6.0.1 yields the following result.

Corollary 5.3.1. *Any node-symmetric network of size N with diameter $D = \Omega(\log N)$ has routing number $\Theta(D)$.*

Since the butterfly is node-symmetric, Corollary 5.3.1 yields that the routing number of a butterfly of size N is $\Theta(\log N)$. As any bounded degree network of size N has a diameter of $\Omega(\log N)$, and any expander has a constant expansion, we obtain the following result together with Theorem 5.4.2, which will be presented in the next section.

Corollary 5.3.2. *Any bounded degree expander of size N has a routing number of $\Theta(\log N)$.*

5.4 Routing Number vs. Expansion

In this section we establish a relationship between the routing number and the expansion of a network. In [LR88] the following result is shown.

Theorem 5.4.1. *Let H be a bounded degree network of size N with expansion α. Given any N-node bounded degree network G, and any 1-1 embedding of the nodes of G onto the nodes of H, the edges of G can be simulated by paths in H with congestion and dilation at least $\Omega(\frac{1}{\alpha})$ and at most $O(\frac{\log N}{\alpha})$.*

Together with Theorem 6.0.1 we obtain the following result.

Theorem 5.4.2. *Any bounded degree network of size N with expansion α has a routing number of at least $\Omega(\frac{1}{\alpha})$ and at most $O(\frac{\log N}{\alpha})$.*

As shown in the next theorem, these bounds are tight.

Theorem 5.4.3. *For all $1/N \leq \alpha \leq 1/\log N$, there exists a bounded degree network of size N with expansion $\Theta(\alpha)$ and routing number $\Theta(\frac{1}{\alpha})$. Furthermore, for all $\log N/N \leq \alpha \leq 1$, there exists a bounded degree network of size N with expansion $\Theta(\alpha)$ and routing number $\Theta(\frac{\log N}{\alpha})$.*

Proof. Let us start with constructing networks for the first case. According to Corollary 5.3.1, the wrap-around butterfly (WBF) of size N has routing number $\Theta(\log N)$. Since for any d-dimensional WBF both of its two $(d-1)$-dimensional sub-butterflies have a size of $d \cdot 2^{d-1}$, but only 2^d edges leaving them, Theorem 5.4.2 yields that the expansion of a WBF of size N is bounded by $\Theta(\frac{1}{\log N})$. If we replace each edge of the WBF by a path of length k, its two $(d-1)$-dimensional sub-butterflies have a size of $\Theta(kd \cdot 2^{d-1})$, but still only 2^d edges leaving them, that is, the expansion of the resulting network is $\Theta(\frac{1}{k \log N})$. On the other hand, it is easy to show (by using, for instance, the random rank protocol together with Valiant's trick, see Section 7.2) that the worst case time for routing a permutation in this network, and hence its routing number is $\Theta(k \log N)$. This yields the first statement of the theorem.

Now we show how to construct networks for the second case. According to Corollary 5.3.2, we know that every bounded degree expander of size N has expansion $\Theta(1)$ and routing number $\Theta(\log N)$. Let G be any of these networks. If we replace each edge of G by a path of length k, it is easy to see that the expansion of the resulting network is $\Theta(\frac{1}{k})$. On the other hand, any permutation routing strategy in this network can be simulated by G in asymptotically the same time (each node in G simulates $k/2$ nodes of each path representing one of its edges). Hence, the worst case time for routing a permutation in this network is at least the worst case time of routing a k-relation in G, and therefore its routing number is $\Omega(k \log N)$. This yields together with Theorem 5.4.2 the second statement of the theorem.   □

Theorem 5.4.3 shows that in general the expansion of a network can only be used to describe its routing number to within a factor of $\log N$. The question therefore is whether there exist simple structural properties of networks that yield a more accurate bound on their routing number.

5.5 Computing Efficient Path Systems

By using the routing number, we showed that path systems exist for arbitrary networks that together with Valiant's trick allow the online construction of efficient path collections for arbitrary permutation routing problems. However, we did not address so far the problem of how difficult it is to compute such a path system. In the following, we will describe a polynomial time algorithm for constructing efficient path systems for arbitrary networks, and present a simple and fast algorithm for constructing efficient path systems for node-symmetric networks.

5.5.1 An Algorithm for Arbitrary Networks

Consider any network G of size N with routing number R. The following algorithm will serve as a basic building block for our algorithm.

For any $L \in \mathbb{N}$, let G_L denote a leveled network of depth L that has N nodes in every level. For each level, let its node set represent the set of all nodes in G. Two nodes v and w in consecutive levels are connected iff v and w represent the same node in G or the nodes represented by v and w are connected in G. In the first case, assume the edge to have infinite capacity, and in the second case, assume the edge to have capacity N. Let all edges be directed downwards. Consider the problem of sending one unit of flow from every node at the top level to every node at the bottom level. If we allow fractional flows then linear programming can be used to find a solution (or stop with the answer that no solution exists) in polynomial time (see Section 1.4.3).

Now we are ready to formulate our algorithm for finding an efficient path system.

1. Find the minimum L for which there is a solution for the multicommodity flow problem for G_L as stated above.
2. Transform the multicommodity flow for G_L into a multicommodity flow for G by identifying nodes in G_L representing the same node in G. Since every edge in G_L representing an edge in G has capacity N, at most a flow of $L \cdot N$ traverses any edge in G.
3. Use Raghavan's method [Ra88] for converting fractional flows into integral flows. This results in a path system for G with dilation at most L and expected congestion at most $L + O(\sqrt{\frac{L \log N}{N}})$.

Clearly, the algorithm runs in polynomial time. It remains to show that $L \leq R$.

For all $i \in [N]$, let S_i denote the optimal schedule for routing permutation $\pi_i : [N] \to [N]$ with $\pi_i(x) = (x + i) \bmod N$. Since the runtime of S_i is at most R, S_i can be easily used to construct a solution to the multicommodity flow problem for G_R, in which every node x in the top row wants to send a unit of flow to node $\pi_i(x)$ in the bottom row, and every edge in G_R that simulates an edge in G has capacity 1. Hence the combination of all these flows results in a solution to our multicommodity flow problem stated above. Since we require L to be minimal, $L \leq R$.

5.5.2 An Algorithm for Node-Symmetric Networks

For node-symmetric networks, there exists a much faster algorithm than the algorithm above. Consider any node-symmetric network G of size N with diameter D. In the proof of Theorem 5.1.2 we have seen that, if for every source-destination pair (u, w) a path from u to w is chosen uniformly at random from all possible shortest paths that connect u with w, then w.h.p. this results in a shortest path system with an expected congestion of $D + O(\sqrt{\frac{D \log N}{N}})$. Let us replace this random experiment by the following strategy for each node pair (u, w), which is much easier to implement:

Start the construction of the shortest path with u. Let v be the current endpoint of the path. From all neighbors of v that can be used to extend the shortest path, choose one uniformly at random, and continue the construction of the path at this neighbor. Repeat this random experiment until w is reached.

Let the $p_{u,w}^v$ and $\mathrm{E}[C_v]$ be defined as in the proof of Theorem 5.1.2. Since G is node-symmetric, there exists for every node v' in G an automorphism φ that maps v to v'. As is easy to see, the uniform choice of a candidate for extending a shortest path ensures that $p_{u,w}^v = p_{\varphi(u),\varphi(w)}^{\varphi(v)}$ for all node pairs (u, w). This yields $\mathrm{E}[C_v] = \mathrm{E}[C_{v'}]$ for all nodes v and v' in G (see the proof of Theorem 5.1.2). Hence, $\mathrm{E}[C_v]$ is the same for every node v in G, namely at most $N \cdot D$. Thus we end up with a shortest path system with expected congestion $\bar{C} = D + O(\sqrt{\frac{D \log N}{N}})$, w.h.p.

5.6 Summary of Main Results

At the end of this chapter we give a summary of its main results. After defining the routing number we showed in Section 5.1 the following result (see Theorem 5.1.1).

For any network of size N with routing number R there is a simple path system with dilation at most R and expected congestion at most R.

Hence *every* network has a simple path system with asymptotically optimal dilation and expected congestion. Using this together with Valiant's trick, we could show the following result in Section 5.2 (see Corollary 5.2.2).

For any network of size N with routing number R there exists a simple path system $\mathcal{P}$, such that routing any h-relation along paths in $\mathcal{P}$ using Valiant's trick has dilation at most $2R$ and congestion at most $2h \cdot R + O(\sqrt{\max\{h \cdot R, \log n\} \cdot \log n})$, w.h.p.

This implies that, if there is an online protocol that can route packets along an arbitrary simple path collection with dilation D and congestion C in time $O(C + D)$ then a combination of the path selection strategy in Corollary 5.2.2 and this protocol would yield an online routing scheme that reaches the optimal routing performance of permutation routing in arbitrary networks. Leighton *et al.* [LMR88, LMR94] have demonstrated that at least offline this is always possible. In Chapter 7 we will show how close oblivious routing protocols can get to this bound.

In Section 5.3, we determined the routing number of arbitrary node-symmetric networks and bounded degree expanders. Moreover, we showed in Section 5.4 that the expansion of a network of size N in general can only be used to describe its routing number to within a factor of $\log N$. Finally, we described in Section 5.5, how efficient path systems can constructed in polynomial time for arbitrary networks.

6. Offline Routing Protocols

In Section 5.2 we saw that for any network G of size N with routing number R there exists a simple path system $\mathcal{P}$, such that routing any permutation along paths in $\mathcal{P}$ using Valiant's trick has dilation at most $2R$ and congestion $2R + O(\sqrt{\max\{R, \log n\}} \cdot \log n)$, w.h.p. On the other hand we saw that, on average, a permutation can not be routed below $\Omega(R)$ steps, whatever protocol used. Hence, in order to get an (asymptotically) optimal average permutation routing time for arbitrary networks, we need a protocol that can route packets along any simple path collection with congestion C and dilation D in $O(C + D)$ steps. In a celebrated paper, Leighton, Maggs and Rao [LMR88] could prove the following result.

Theorem 6.0.1. *For any set of packets with simple paths having congestion C and dilation D, there is an offline schedule that needs $O(C + D)$ steps to route all packets, using buffers of constant size.*

The proof of this theorem can be found in [LMR88] (see also [LMR94]). In [LMR96] it is furthermore shown that there is an algorithm that computes such a schedule in time $O(P \cdot \log \log P \cdot \log P)$ steps, where P is the sum of the lengths of the paths.

The proof Leighton *et al.* used, however, has several disadvantages: It is very difficult to understand, and it makes no attempt to reduce the constants hidden in the runtime bound and the bound for the buffer size. (In fact, the constants hidden are beyond one million.)

In this chapter, we present alternative proofs for the existence of offline protocols with runtime $O(C + D)$, that are much easier to understand. The first two aim at keeping the runtime small, and the third aims at keeping the buffer size small. As in [LMR88], the basic argument used in the proofs is the Lovász Local Lemma (see [AES92], p.55).

Lemma 6.0.1 (Lovász Local Lemma). *Let $A_1, \ldots, A_n$ be a set of "bad" events in an arbitrary probability space. Suppose that each event A_i is mutually independent of all other events A_j but at most b, and that $\Pr[A_i] \leq p$. If $ep(b+1) < 1$ then, with probability greater than zero, no bad event occurs.*

Besides presenting proofs for efficient offline protocols, we also show how these protocols can be applied to network simulations. In particular, we present techniques that will be used in Chapter 8 and Chapter 9 to construct fast deterministic and compact routing strategies.

6.1 Keeping the Routing Time Low

In this section we present two proofs that aim at minimizing the runtime of offline protocols. For arbitrary C and D, the following result holds.

Theorem 6.1.1. *For any set of packets with simple paths having congestion C and dilation D, there is an offline schedule that needs at most $39(C + D)$ time to route all packets, using buffers of size $O(\log^3(C + D))$.*

Proof. Our strategy for constructing an efficient schedule is to make a succession of refinements, starting with an initial schedule S_0 in which each packet is forwarded at every time step until it reaches its final destination. This initial schedule is as short as possible; its length is only D. Unfortunately, as many as C packets may have to use an edge at a single time step in S_0, whereas in the final schedule at most one packet is allowed to use an edge at each step. Each refinement will bring us closer to meeting this requirement by bounding the congestion within smaller and smaller time intervals. The proof uses the Chernoff bounds and the Lovász Local Lemma at each refinement step. In the following, a *t-interval* denotes a time interval (i.e., a sequence of consecutive time steps) of length t. A *t-frame* is defined as a t-interval that starts at an integer multiple of t.

Let $I_0 = \max\{C, D\}$ and $I_j = \log I_{j-1}$ for all $j \geq 1$. The first step is to assign an initial delay to each packet, chosen independently and uniformly at random from the range $\Delta_1 = [C]$. In the resulting schedule, S_1, a packet that is assigned a delay of δ waits in its source node for δ steps, then moves on without waiting again until it reaches its destination. The length of S_1 is at most $D + C$. We use the Lovász Local Lemma to show that if the delays are chosen independently and uniformly at random and I_1 is sufficiently large, then with nonzero probability the congestion at any edge in any I_1^3-interval is at most $(1 + \frac{3}{I_1})I_1^3$. Thus, such a set of delays must exist.

To apply the Lovász Local Lemma, we associate a bad event with each edge. The bad event for edge e is that more than $C_1 = (1 + \frac{3}{I_1})I_1^3$ packets use e in some I_1^3-interval. For this we have to bound the dependence b among the bad events and the probability p that a bad event occurs.

We first bound the dependence. Whether or not a bad event occurs depends solely on the delays assigned to the packets that pass through the corresponding edge. Thus, two bad events are independent unless some packet passes through both of the corresponding edges. Since at most C packets pass through an edge and each of these packets passes through at most D other edges, the dependence b of the bad events is at most $C \cdot D$.

Next we bound the probability that a bad event occurs. Let us consider some fixed edge e and I_1^3-interval J. Let the packets traversing e be numbered from 1 to $m \leq C$. For every $i \in \{1, \ldots, m\}$, let the binary random variable X_i be one if and only if packet i traverses e during J. Let $X = \sum_{i=1}^{m} X_i$. Since the packets choose their delays uniformly at random from a range of size C, we get $\Pr[X_i = 1] \leq I_1^3/C$ for all $i \in \{1, \ldots, m\}$. Hence,

$$E[X] = \sum_{i=1}^{m} E[X_i] \leq I_1^3 .$$

Together with the Chernoff bounds we therefore get with $\epsilon = \frac{3}{I_1}$ that

$$\Pr[X \geq C_1] = \Pr[X \geq (1+\epsilon)I_1^3] \leq e^{-\left(\frac{3}{I_1}\right)^2 I_1^3/2} \leq 2^{-5I_1} .$$

Since for each edge there are at most $D + C$ I_1^3-intervals to consider, the probability that a bad event occurs for edge e is bounded by

$$p \leq (D+C)2^{-5I_1} < \frac{1}{e(CD+1)}$$

for all $I_1 \geq 4$, since $32 \cdot e(256 + 1) < 2^{-20}$. Thus the product $ep(b + 1)$ is less than 1 for all $I_1 \geq 4$ and therefore, by the Lovász Local Lemma, there is some assignment of delays such that the congestion in each I_1^3-interval is bounded by $(1 + \frac{3}{I_1})I_1^3$.

We now break schedule S_1 into I_1^4-frames and continue to schedule each frame independently. So each frame can be viewed as a separate scheduling problem where the origin of a packet is its location at the beginning of the frame, and its destination is its location at the end of the frame. Our next refinement step will be to choose, for each frame, a random initial delay for each packet that visits at least one edge within this frame. In the resulting schedule S_2, the frames (enlarged by their delay ranges) are executed one after the other in a way that a packet that is assigned a delay δ in some frame F waits at its first edge in F for (additional) δ steps, and then moves on without waiting until it traverses its last edge in F.

Let us concentrate in the following on some fixed frame F. Let each packet choose an additional delay out of the range $\Delta_2 = [I_1^3 - I_2^3]$. Hence the length of the resulting schedule for this frame is at most $I_1^4 + I_1^3$. We use the Lovász Local Lemma to show that if the delays are chosen independently and uniformly at random and I_2 is sufficiently large, then with nonzero probability the congestion in any I_2^3-interval is at most

$$C_2 = \left(1 + \frac{3}{I_1}\right)\left(1 + \frac{4}{I_2}\right)\left(\frac{1}{1 - (I_2/I_1)^3}\right) I_2^3 .$$

We again associate a bad event with each edge. The bad event for edge e is that more than C_2 packets use e in some I_2^3-interval.

We first bound the dependence of these events. Whether or not a bad event occurs solely depends on the delays assigned to the packets that pass through the corresponding edge. Since at most $C_1 \cdot I_1$ packets pass through an edge in any frame and each of these packets passes through at most I_1^4 other edges, the dependence of the bad events is at most $C_1 \cdot I_1^5$.

Next we bound the probability that a bad event occurs. Let us consider some fixed edge e and I_2^3-interval J. We know according to the first refinement step that every I_1^3-interval has congestion at most C_1. Since the packets can only choose delays up to $I_1^3 - I_2^3$ and J covers I_2^3 time steps, at most C_1 packets are able to traverse this edge during J. Let these packets be numbered from 1 to $m \leq C_1$. For every $i \in \{1, \ldots, m\}$, let the binary random variable X_i be one if and only if packet i traverses e during J. Let $X = \sum_{i=1}^m X_i$. Since the packets choose their delays uniformly at random from a range of size $I_1^3 - I_2^3$, we get $\Pr[X_i = 1] \leq I_2^3/(I_1^3 - I_2^3)$ for all $i \in \{1, \ldots, m\}$. Hence,

$$E[X] = \sum_{i=1}^m E[X_i] \leq \left(1 + \frac{3}{I_1}\right) \left(\frac{1}{1 - (I_2/I_1)^3}\right) I_2^3 .$$

Together with the Chernoff bounds we therefore get with $\epsilon = \frac{4}{I_2}$ and $\alpha_2 = (1 + \frac{3}{I_1})(\frac{1}{1-(I_2/I_1)^3})$ that

$$\Pr[X \geq C_2] = \Pr[X \geq (1+\epsilon)\alpha_2 I_2^3] \leq e^{-\left(\frac{4}{I_2}\right)^2 \alpha_2 I_2^3/2} \leq e^{-8\alpha_2 I_2/2} .$$

Since for each edge there are at most $I_1^4 + I_1^3$ I_2^3-intervals to consider, the probability that a bad event occurs for edge e is bounded by

$$p \leq (I_1^4 + I_1^3)e^{-8\alpha_2 I_2} < \frac{1}{e(C_1 I_1^5 + 1)}$$

for all $I_2 \geq 4$, since $(2^{16} + 2^{12})e((1 + \frac{3}{16})2^{32} + 1)e^{-8 \cdot 1.1875 \cdot 4} < 1$. Thus the product $ep(b+1)$ is less than 1 for all $I_2 \geq 4$ and therefore, by the Lovász Local Lemma, there is some assignment of delays such that the congestion in each I_2^3-interval is bounded by C_2.

We continue to refine each I_2^4-frame in F recursively until we reach a round k, in which $I_k = 4$. (Note that if there is no I_k with $I_k = 4$ then we use a slightly different refinement scheme with I_j's instead of I_js. In almost all cases, the I_j's can be defined as $I_0' = I_0$, $I_1' = I_1$, $I_2' = \max\{4, s\}$ with $\log^* s = \lfloor \log^* I_0 \rfloor$, and $I_{j+1}' = \log I_j'$ for all $j \geq 2$ to obtain (up to very small terms) the same factors as below. In this case we have $I_k' = 4$ for some k. For the remaining cases (such as I_0 is small) it is also easy to find suitable I_j's.) We end up with a schedule S_k with total length at most

$$(D+C) \prod_{j=1}^{k-1} \left(1 + \frac{1}{I_j}\right) \leq 1.1(D+C)$$

and with a congestion C_k in each I_k^3-interval of at most

$$I_k^3 \left(1 + \frac{3}{I_1}\right) \prod_{j=1}^{k-1} \left(1 + \frac{4}{I_{j+1}}\right) \left(\frac{1}{1 - (I_{j+1}/I_j)^3}\right) \leq 2.5 I_k^3 \ .$$

It remains to show how to get down to 1-frames. Let us break schedule S_k into frames of size $4^3 = 64$. Then it remains to show for a single frame, how to refine it to 1-frames. Consider some fixed frame F. Let us perform the random experiment of assigning to each packet in F an additional delay chosen uniformly and independently at random from $\Delta = [64]$. Then the average number of packets traversing an edge at each time step is bounded by 2.5. Let the random variable X denote the number of packets traversing an edge e at some time step. Together with the Chernoff bounds we therefore get with $\epsilon = 6$ that

$$\Pr[X \geq (1 + \epsilon)2.5] \leq \left(\frac{e^6}{7^7}\right)^{2.5} \ .$$

Since there are at most $2 \cdot 64$ time steps to consider, the probability p that at least $(1 + \epsilon)2.5$ packets traverse e at some time step is bounded by

$$2 \cdot 64 \left(\frac{e^6}{7^7}\right)^{2.5} < \frac{1}{e(2.5 \cdot 64^2 + 1)} \ .$$

Since the dependence among the edges is at most $2.5 \cdot 64^2$, the Lovász Local Lemma yields that delays can be assigned to the packets such that no more than $(1 + 6)2.5$ packets cross an edge at each time step.

Putting all refined frames together and simulating the traversal of at most $7 \cdot 2.5$ packets per time step over one link by $7 \cdot 2.5$ time steps results in an offline schedule that needs at most $1.1 \cdot 2 \cdot 7 \cdot 2.5(C + D) \leq 39(C + D)$ time steps to route all packets along their paths.   ☐

With a more involved proof, in which the nonuniform version of the Lovász Local Lemma (see [AES92]) is used to bound both the congestion within certain frames and the contention (i.e., the maximum number of packets crossing an edge at the same time), the buffer size in Theorem 6.1.1 can be reduced to $O(\log(C + D))$.

The question is, what the real constant in front of the $(C + D)$ is for a fastest possible offline protocol. If $C \gg D$, then it is easy to show that there exists an offline schedule with routing time at most $(1 + \epsilon)C$ for some $0 < \epsilon < 1$ depending on C and D.

Theorem 6.1.2. *For every simple path collection with dilation D and congestion $C \geq 6D \log C$ there is an offline protocol that needs*

$$\left(1 + 3\sqrt[3]{\frac{6D \log C}{C}}\right) C$$

steps to route all packets, using buffers of size at most $2C/D$.

Proof. Our strategy here for constructing an efficient schedule is to make just one refinement to the initial schedule S_0 which is defined as in the previous proof. Let us perform the random experiment of assigning to each packet a delay chosen uniformly and independently at random from $\Delta = [\delta D]$, where $\delta = \sqrt[3]{C/(6D \log C)}$. Then the expected number of packets that want to traverse any edge at any time step is at most $C/\delta D$. Let X denote the expected contention at some edge e at some fixed time step. Since the packets choose their delays independently, we can use the Chernoff bounds to get that for $\epsilon = 1/\delta$

$$\Pr\left[X \geq (1+\epsilon)\tfrac{C}{\delta D}\right] \leq e^{-\epsilon^2 C/(2\delta D)} = e^{-3\log C} \ .$$

Since there are at most $(1 + \delta)D$ time steps to consider, the probability p that at least $(1+\epsilon)C/\delta D$ packets traverse e at some time step is bounded by

$$(1+\delta)De^{-3\log C} < \frac{1}{e(C \cdot D + 1)}$$

Since the dependence among the edges is at most $C \cdot D$, the Lovász Local Lemma yields that delays can be assigned to the packets such that less than $(1 + \epsilon)C/\delta D$ packets cross an edge at each time step. Simulating each time step of this schedule by $(1 + \epsilon)C/\delta D$ time steps yields a schedule in which at most one packet traverses each link in one time step, that has runtime at most

$$(1+\epsilon)\tfrac{C}{\delta D}(1+\delta)D = \left(1 + \frac{1}{\delta}\right)^2 C \leq \left(1 + \frac{3}{\delta}\right)C \ ,$$

using buffers of size at most $(1 + \epsilon)C/\delta D$. $\square$

One could ask whether it is possible to find algorithms that compute a schedule with a smaller or even optimal delay. However, it has been shown by Di Ianni [Di96] that in general the problem of minimizing the delay beyond some additive error of D^ϵ of routing packets along a simple path collection can not be solved in polynomial time unless $P = NP$.

6.2 Keeping the Buffer Size Low

In this section we present an offline protocol that only requires edge buffers of size three. Hence this protocol can even be used in systems with severe hardware restrictions. The constant hidden in the time bound of the offline protocol, however, is very high.

Theorem 6.2.1. *For any set of packets with simple paths having congestion C and dilation D, there is an offline schedule for buffers of size 3 that needs time $O(C + D)$.*

Proof. Our strategy for constructing an efficient schedule is again to make a succession of refinements, starting with an initial schedule S_0 in which each packet is forwarded at every time step until it reaches its final destination. New in this proof is that we use so-called *secure edges*. Given any schedule S, an edge along the path of a packet P is defined as *secure* in some time interval I if no other packet except P waits there at any time step during I in S. We will show how these edges will enable us to route with buffer size three.

Let a *t-interval* again be defined as a time interval of length t. Let $I_0 = \max\{C, D\}$ and $I_j = \log I_{j-1}$ for all $j \geq 1$. The first step is to assign an initial delay to each packet, chosen independently and uniformly at random from the range $\Delta_1 = [2C]$. In the resulting schedule, S_1, a packet that is assigned a delay of δ waits in its source node for δ steps, then moves on without waiting again until it reaches its destination. The length of S_1 is at most $D + 2C$. We use the Lovász Local Lemma to show that if the delays are chosen independently and uniformly at random and I_1 is sufficiently large, then with nonzero probability the congestion at any edge in any I_1^2-interval is at most $C_1 = \frac{1}{2}(1 + \frac{4}{\sqrt{I_1}})I_1^2$. Thus, such a set of delays must exist.

To apply the Lovász Local Lemma, we associate a bad event with each edge. The bad event for an edge is that the condition above is not fulfilled.

We first bound the dependence. Whether or not a bad event occurs depends solely on the delays assigned to the packets that pass through the corresponding edge. Thus, two bad events are independent unless some packet passes through both of the corresponding edges. Since at most C packets pass through an edge and each of these packets passes through at most D other edges, the dependence b of the bad events is at most $C \cdot D$.

Next we bound the probability that a bad event occurs. Consider some fixed edge e and I_1^2-interval J. Let the packets traversing e be numbered from 1 to $m \leq C$. For every $i \in \{1, \ldots, m\}$, let the binary random variable X_i be one if and only if packet i traverses e during J. Let $X = \sum_{i=1}^{m} X_i$. Since the packets choose their delays uniformly at random from a range of size $2C$, we get $\Pr[X_i = 1] \leq I_1^2/2C$ for all $i \in \{1, \ldots, m\}$. Hence,

$$E[X] = \sum_{i=1}^{m} E[X_i] \leq \frac{I_1^2}{2} .$$

Together with the Chernoff bounds we therefore get with $\epsilon = \frac{4}{\sqrt{I_1}}$ and $I_1 \geq 16$ that

$$\Pr[X \geq C_1] = \Pr[X \geq (1 + \epsilon)I_1^2/2] \leq e^{-\epsilon^2 I_1^2/4} \leq e^{-4I_1} .$$

Since for each edge there are at most $D + 2C$ I_1^2-intervals to consider, the probability p that a bad event occurs for edge e is bounded by

$$p \leq (D + 2C)e^{-4I_1} < \frac{1}{e(CD + 1)} .$$

Thus the product $ep(b + 1)$ is less than 1 and therefore, by the Lovász Local Lemma, there is some assignment of delays such that the congestion in each I_1^2-interval is bounded by C_1. Let the corresponding schedule be called S_1.

Before we show how to refine S_1 we need some notation. An I_1^2-interval in S_1 is called *waiting zone* if it starts at an integer multiple of I_1^7. Given a waiting zone Z, the path traversed by a packet within Z is called its Z-path. Our aim is to choose a suitable edge in every packet's Z-path, called its *waiting edge*. A waiting edge on a packet's Z-path is called *secure* if no other Z-path chooses it as its waiting edge. Since the source and destination of a packet can be seen as a secure waiting edge, we only consider the problem of finding secure waiting edges for all packets that do not start or end to travel within Z. This ensures that whenever we consider a Z-path of a packet, it has length I_1^2. The following lemma shows that under this assumption an assignment of secure waiting edges exists for all Z-paths.

Lemma 6.2.1. *For any collection of paths of length d with congestion at most d there is an assignment of edges to paths such that every path contains an edge assigned to it and every edge is assigned to at most one path.*

Proof. Let $G = (V_1, V_2, E)$ be a bipartite graph with V_1 representing the paths and V_2 representing all edges used by these paths. A node $u \in V_1$ is connected to node $v \in V_2$ if the path represented by u contains the edge represented by v. Since an edge is used by at most d paths, all nodes in V_2 have a degree of at most d. As all nodes in V_1 have a degree of d, it holds for every subset $U \subseteq V_1$ that $|\Gamma(U)| \geq |U|$. Otherwise there must exist a node in $\Gamma(U)$ with degree larger than d. From Theorem 3.3.1 it follows that there must exist a matching of size $|V_1|$ in G. Thus for each path an edge can be chosen such that every edge is assigned to at most one path. $\square$

We now break S_1 into subschedules $S_{1,i}$ of length $I_1^7 + I_1^2$ at the secure edges as shown in Figure 6.1, and continue to refine each subschedule independently.

Thus let us concentrate in the following on some fixed subschedule $S_{1,i}$. Let each packet in $S_{1,i}$ choose an (additional) initial delay uniformly at random out of the range $\Delta_2 = [I_2^2 - I_2^2]$. We use the Lovász Local Lemma to show that if the delays are chosen independently and uniformly at random and I_2 is sufficiently large, then with nonzero probability the congestion in any I_2^2-interval is at most

$$C_2 = \frac{1}{2}\left(1 + \frac{4}{\sqrt{I_1}}\right)\left(1 + \frac{8}{\sqrt{I_2}}\right)\left(\frac{1}{1 - (I_2/I_1)^2}\right) I_2^2 .$$

We again associate a bad event with each edge. The bad event for edge e is that more than C_2 packets use e in some I_2^2-interval.

We first bound the dependence of these events. Whether or not a bad event occurs solely depends on the delays assigned to the packets that pass through

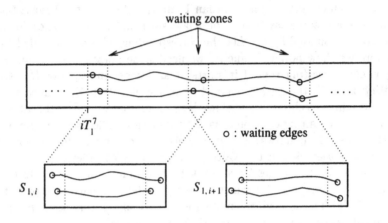

Fig. 6.1. Splitting S_1 into subschedules $S_{1,i}$.

the corresponding edge. Since at most $C_1(I_1^5 + 1)$ packets pass through an edge in $S_{1,i}$ and each of these packets passes through at most $I_1^7 + I_1^2$ other edges, the dependence of the bad events is at most $2C_1 \cdot I_1^{12}$.

Next we bound the probability that a bad event occurs. Let us consider some fixed edge e and I_2^2-interval J. We know according to the first refinement step that every I_1^2-interval has congestion at most C_1. Since the packets can only choose delays up to $I_1^2 - I_2^2$ and J covers I_2^2 time steps, at most C_1 packets are able to traverse this edge during J. Let these packets be numbered from 1 to $m \leq C_1$. For every $i \in \{1, \ldots, m\}$, let the binary random variable X_i be one if and only if packet i traverses e during J. Let $X = \sum_{i=1}^{m} X_i$. Since the packets choose their delays uniformly at random from a range of size $I_1^2 - I_2^2$, we get $\Pr[X_i = 1] \leq I_2^2/(I_1^2 - I_2^2)$ for all $i \in \{1, \ldots, m\}$. Hence,

$$\mathrm{E}[X] = \sum_{i=1}^{m} \mathrm{E}[X_i] \leq \frac{1}{2}\left(1 + \frac{4}{\sqrt{I_1}}\right)\left(\frac{1}{1 - (I_2/I_1)^2}\right) I_2^2 .$$

Together with the Chernoff bounds we therefore get with $\epsilon = \frac{8}{\sqrt{I_2}}$, $I_2 \geq 64$ and $\alpha_2 = \frac{1}{2}(1 + \frac{4}{\sqrt{I_1}})(\frac{1}{1-(I_2/I_1)^2})$ that

$$\Pr[X \geq C_2] = \Pr[X \geq (1 + \epsilon)\alpha_2 I_2^2] \leq \mathrm{e}^{-\epsilon^2 \alpha_2 I_2^2/2} \leq \mathrm{e}^{-16I_2} .$$

Since for each edge there are at most $I_1^7 + 2I_1^2$ I_2^2-intervals to consider, the probability that a bad event occurs for edge e is bounded by

$$p \leq (I_1^7 + 3I_1^2)\mathrm{e}^{-16I_2} < \frac{1}{\mathrm{e}(2C_1 I_1^{12} + 1)} .$$

Thus the product $ep(b+1)$ is less than 1 and therefore, by the Lovász Local Lemma, there is some assignment of delays such that the congestion in each I_2^2-interval is bounded by C_2. Let us call the resulting schedule $S'_{1,i}$.

inner region

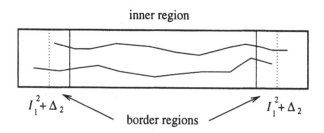

$I_1^2 + \Delta_2$ $I_1^2 + \Delta_2$

border regions

Fig. 6.2. Separating $S'_{1,i}$ into three regions.

Next, we separate $S'_{1,i}$ into three regions as shown in Figure 6.2. Let the first $2I_1^2 - I_2^2$ and the last $2I_1^2 - I_2^2$ time steps of $S'_{1,i}$ be called *border region* and the rest *inner region*. Clearly, all paths traversed in $S'_{1,i}$ that do not represent the first or last piece of a path of the original path collection start and end in the border regions. We consider as a waiting zone all I_2^2-intervals that start in the inner region of $S'_{1,i}$ at an integer multiple of I_2^7 (counting from the beginning of the inner region). In case that the last waiting zone is more than I_2^2 steps away from the beginning of the border region at the end of $S'_{1,i}$, we also declare the last I_2^2 steps before that border region as waiting zone. Given a waiting zone Z, our aim is to choose as above a secure waiting edge in every packet's Z-path. Since the source and destination of a packet can be seen as a secure waiting edge, we only need to consider packets that neither start nor end in Z. Hence, all Z-paths to be considered have length I_2^2. For large enough I_2, the congestion C_2 in each of the waiting zones is at most I_2^2. Hence Lemma 6.2.1 implies that there exists an assignment of secure waiting edges to all Z-paths.

Break now $S'_{1,i}$ into three pieces at the secure edges of the first and the last waiting zone. Since both the congestion and the length of the longest path in the first piece is at most $2I_1^2$, according to Theorem 3.3.2 the paths can be separated into at most $4I_1^4 + 1$ sets of disjoint paths. Scheduling each of these sets one after the other takes at most $2I_1^2(4I_1^4 + 1)$ time steps. Because of our choice of secure edges for the first and second refinement step, this only requires buffers of size three (one slot for a packet that has not been sent yet, one slot for traveling packets, and one slot for a packet that has already been sent). Using the same strategy for the last piece, we obtain a runtime of $O(I_1^6)$ for scheduling the first and the last piece of $S'_{1,i}$. Let us call the resulting schedule $S''_{1,i}$. Processing the $S''_{1,i}$ one after the other, we obtain schedule S_2.

In order to refine S_2 to S_3, we break the inner regions of all the subschedules $S''_{1,i}$ at the secure edges into subschedules $S_{2,j}$ of length at most $I_2^7 + I_2^2$. Each of these subschedules is then refined in the same way as $S_{1,i}$ before. We continue with the refinements until we reach a stage k in which

$$\frac{1}{2}\left(1+\frac{4}{\sqrt{I_1}}\right)\left[\prod_{j=2}^{k}\left(1+\frac{8}{\sqrt{I_j}}\right)\left(\frac{1}{1-(I_j/I_{j-1}^2)}\right)\right]I_k^2>I_k^2 \ .$$

In this case, I_k is only a constant. Separating the paths that remain to be scheduled in the inner regions into sets of disjoint paths and scheduling them one after the other then yields a schedule S_k with runtime at most

$$(D+2C)\left(\left(\sum_{j=1}^{k-1}\frac{1}{I_j^7}\cdot O(I_j^6)\right)+O(1)\right)=O(C+D)$$

that requires only buffers of size three.   □

It is easy to see that we can reduce the buffer size to two without changing the time bound if we allow that positions of packets can be exchanged when moving packets. A further reduction to a buffer size of one, however, might be difficult to achieve together with a runtime of $O(C+D)$.

6.3 Applications

In the following we show how the offline protocols can be used for efficient network emulations. We start with describing a simple emulation strategy using 1–1 embeddings. Afterwards, we describe, how to obtain efficient emulations using 1–many embedding strategies. Especially the 1–many embedding strategy will be used later to obtain fast deterministic and compact routing protocols.

6.3.1 Network Emulations Using 1–1 Embeddings

Consider the problem of simulating an arbitrary routing step of a network G of size N by a network H of size N. Let us embed the nodes of G 1–1 into the nodes of H in an arbitrary way. Clearly, any routing step in G can be extended to the situation that along every edge in G a packet has to be sent. Hence, in order to simulate an arbitrary routing step in G by H, we need

– a path collection in H such that for every edge $\{u,v\}$ in G there is a path in H that connects the two nodes simulating u and v, and
– a protocol that sends a packet along each of these paths in an efficient way.

Let G have degree d. Then the edges of G can be colored by $2d$ colors such that no two adjacent edges use the same color (apply Theorem 3.3.2). Let R be the routing number of H. According to the definition of the routing number, there exists for each color a path collection for simulating edges of that color in G by H with dilation and congestion at most R (see Remark 5.0.1). Hence altogether there exists a path collection for simulating the edges in G by H

with dilation at most R and congestion at most $2d \cdot R$. Applying the offline protocol presented in Theorem 6.2.1 for sending one packet along each of these paths yields the following result.

Theorem 6.3.1. *Let G be an arbitrary network of size N and degree d, and H be an arbitrary network of size N with routing number R. Then any routing step of G can be simulated by H in time $O(d \cdot R)$, using only constant size buffers.*

If we choose G to be a multibutterfly with constant degree, we arrive at the following result.

Theorem 6.3.2. *Let H be an arbitrary network of size N with routing number R, for which there exists a multibutterfly of equal size. Then an online deterministic routing strategy can be constructed for routing any permutation in H in $O(R \log N)$ steps, using only constant size buffers.*

Proof. According to [Up92], there exist multibutterfly networks with constant degree, such that any global permutation can be routed in $O(\log N)$ time using constant size buffers, where N is the size of the respective multibutterfly. Using this together with the strategy described in Theorem 6.3.1 for simulating an arbitrary routing step in the multibutterfly yields the time bound of Theorem 6.3.2.   □

In Chapter 8 we show that multibutterflies can be constructed such that for any $N \geq 1$ there exists a multibutterfly of size approximately N.

6.3.2 Network Emulations Using 1–Many Embeddings

Consider the problem of simulating one routing step of an arbitrary network G by a network H for the case that the size of G is smaller than the size of H. We start with describing how to embed G in H. In order to simplify the construction, let H be a network of size N, and G be a network of size M with at most $N/2$ edges. Let $d_1, \ldots, d_M$ be the degree sequence of G, i.e., d_i is the degree of node i in G. Then $\sum_{i=1}^{M} d_i \leq N$. Our strategy is to partition the nodes of H into clusters $C_1, \ldots, C_M$ such that for all $i \in \{1, \ldots, M\}$ cluster C_i consists of d_i nodes representing d_i copies of node i in G. For this we choose an arbitrary spanning tree T in H. Let r be an arbitrary node in T. We mark the nodes in T with numbers in $\{1, \ldots, M\}$ starting with r by calling Mark($1, true, r$):

Algorithm Mark(i, f, v):

i: number of nodes already marked
f: boolean variable indicating whether father has been marked
v: actual node to be considered

if $f = false$ **then**
    mark v with the number ℓ obeying $\sum_{j=1}^{\ell-1} d_j < i \leq \sum_{j=1}^{\ell} d_j$
    set $i = i + 1$
    for every son w of v: call Mark($i, true, w$)
else
    for every son w of v: call Mark($i, false, w$)
    mark v with the number ℓ obeying $\sum_{j=1}^{\ell-1} d_j < i \leq \sum_{j=1}^{\ell} d_j$
    set $i = i + 1$
return the value of i

Basically, the algorithm ensures that on a pass downwards through the tree only every second node is marked such that afterwards on a pass upwards the other half of the nodes can be marked. Consider two nodes marked as i-th and $(i+1)$-st node. Then on a pass downwards, the cases shown in Figure 6.3 can occur. On a pass upwards, the cases shown in Figure 6.4 can occur.

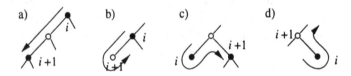

Fig. 6.3. Alternatives for a pass downwards.

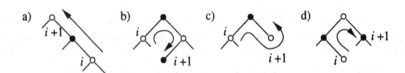

Fig. 6.4. Alternatives for a pass upwards.

As can be seen, the worst case that can happen with this strategy is that two consecutively marked nodes have a distance of 3 (by changing from one subtree into another). Hence nodes belonging to a cluster i have at most a distance of $3d_i$ from each other. Let the nodes of each cluster be connected by an Euler tour along edges in T. (An Euler tour in a tree is defined as a directed cycle that uses any edge in T at most once in every direction.) Then a link is only used by two Euler tours: the Euler tour belonging to that cluster that was built while traversing that link downwards, and the Euler tour belonging to that cluster that was built while traversing that link upwards. Hence the following two results hold.

a) the Euler tour of cluster i has length at most $6d_i$ for all $i \in \{1, \ldots, M\}$, and

b) the maximal number of Euler tours that share the same link is two.

We further simulate every edge in G by a path in H that connects the nodes simulating its endpoints. Let R be the routing number of H. Since our clustering allows the endpoints of edges in G to be distributed in H such that every node in H has to simulate at most one endpoint, there is a path collection in H for simulating the edges in G with congestion at most R and dilation at most R (see Remark 5.0.1).

Consider now the problem of simulating an arbitrary routing step in G. Clearly, any routing step can be extended to the situation that along every edge in G a packet has to be sent. This event can be simulated in the following way by H.

- Moving the packets to the nodes simulating the endpoints of the edges they want to use in G: This can be done by sending the packets along an Euler tour connecting the nodes of the respective cluster in T. Because of (b) this can be coordinated among the clusters deterministically in time $O(\max_i d_i)$ using only constant size buffers.
- Moving the packets along an edge in G: This can be done by sending the packets along the paths simulating edges in G. Since these paths have congestion at most R and dilation at most R, this can be done deterministically in time $O(R)$ using only constant size buffers (see Theorem 6.2.1).

If we restrict the maximum degree in G to be $O(R)$, we get the following result.

Theorem 6.3.3. *Any network H of size N with routing number R can simulate any routing step in a network G with degree $O(R)$ and $O(N)$ edges in $O(R)$ steps, using constant size buffers.*

Hence for networks G of size smaller than the size of H this simulation strategy is better than the strategy used in Theorem 6.3.1.

6.4 Summary of Main Results

In the following, let us summarize the most important results of this chapter. We presented three results about offline protocols. The first two aimed at minimizing the routing time (see Theorems 6.1.1 and 6.1.2):

For any set of packets with simple paths having congestion C and dilation D, there is an offline schedule that needs at most $39(C + D)$ time steps to route all packets, using buffers of size $O(\log^3(C + D))$.

For any set of packets with simple paths having dilation D and congestion $C \geq 6D \log C$, there is an offline schedule that needs at most

$$\left(1 + 3\sqrt[3]{\frac{6D\log C}{C}}\right) C$$

time steps to route all packets, using buffers of size at most $2C/D$.

The third result we presented aimed at minimizing the buffer size (see Theorem 6.2.1).

For any set of packets with simple paths having congestion C and dilation D, there is an offline schedule for buffers of size 3 that needs time $O(C + D)$ to route all packets.

This result can be used to prove the following two results about network emulations (see Theorems 6.3.1 and 6.3.3).

Let G be an arbitrary network of size N and degree d, and H be an arbitrary network of size N with routing number R. Then any routing step of G can be simulated by H in time $O(d \cdot R)$, using constant size buffers.

Any network H of size N with routing number R can simulate any routing step in a network G with degree $O(R)$ and $O(N)$ edges in $O(R)$ steps, using constant size buffers.

Especially the second result will have applications for efficient deterministic and/or compact routing schemes for arbitrary networks. Now let us turn to online protocols.

7. Oblivious Routing Protocols

In this chapter we present several universal oblivious routing protocols that are close to optimal for certain classes of path collections, ranging from the class of leveled path collections to the class of arbitrary (even non-simple) path collections. For each of these protocols, we present its structure and runtime, describe how it can be applied to routing in specific networks, and show its limitations.

In Chapter 5 we have seen that oblivious routing protocols can be used to achieve an efficiency that is close to the routing number, because the dilation and expected congestion of a suitably chosen path system are at most the routing number of the underlying network. Whenever it makes sense, we will therefore use the routing number to describe the runtime of applications of the oblivious routing protocols presented below to specific networks.

7.1 The Random Delay Protocol

The oldest online protocol that deviates only by a factor logarithmic in n, C and D from a best possible runtime of $O(C+D)$ for arbitrary path collections is the protocol presented by Leighton, Maggs and Rao in [LMR88]. We present an extension of it, called here the *random delay protocol*, that can route packets along an arbitrary simple path collection of size n with congestion C and dilation D in $O(C + D \log n)$ steps, w.h.p. (the protocol in [LMR88] requires $O(C + D \log(nD))$ time steps, w.h.p.).

7.1.1 Description of the Protocol

The protocol assumes that all links have bandwidth B (fixed later), that is, up to B packets can traverse a link at one time step. Clearly, each time step for links with bandwidth B can be simulated in B time steps by links with bandwidth 1. The following algorithm is used as a basic building block for the random delay protocol.

Algorithm Route(ℓ):
Each packet is assigned an initial delay, chosen uniformly and independently at random from the range $[C/\log n]$. A packet that is assigned a delay of δ waits in its initial buffer for δ steps and then moves on without waiting again until it reaches its destination or traversed ℓ links. If more than B packets want to use the same link at the same time then all of them stop.

The random delay protocol works as follows.

    repeat
        execute Route(min$\{D, n\}$)
    until all packets reached their destinations

Theorem 7.1.1. *Suppose we are given an arbitrary simple path collection $\mathcal{P}$ of size n with congestion C and dilation D. Then the random delay protocol needs at most $O(C + D \log n)$ time steps to finish routing in $\mathcal{P}$, w.h.p.*

Proof. Let us consider some fixed edge e and time step t during the execution of Route(ℓ). Since at most C packets want to traverse e and each of these packets chooses an initial delay independently at random from a range of size $C/\log n$, the probability that at least $B = \max\{\alpha + 2, 2e\} \log n + 2$ packets want to traverse e at time step t is at most

$$\binom{C}{B} \left(\frac{1}{C/\log n}\right)^B \leq \left(\frac{e \log n}{B}\right)^B \leq \left(\frac{1}{2}\right)^{(\alpha+2)\log n + 2} = \frac{1}{4n^{\alpha+2}} \ .$$

Let us say that a packet P *fails* at edge e if at least B other packets want to use e at the same time as P. Then the probability that P fails at least $k = \lceil D/n \rceil$ times during the execution of the random delay protocol is bounded by

$$\binom{D+k}{k} \left(\frac{1}{4n^{\alpha+2}}\right)^k \leq \left(\frac{4D}{k}\right)^k \left(\frac{1}{4n^{\alpha+2}}\right)^k \leq \left(\frac{1}{n^{\alpha+1}}\right)^k \leq \frac{1}{n^{\alpha+1}} \ .$$

Since there are n packets to consider, the probability that there exists a packet with at least k failures is at most $n \cdot n^{-\alpha-1} = n^{-\alpha}$. Hence, w.h.p. the random delay protocol successfully routes all packets along the given path collection in time

$$B \cdot (\ell + C/\log n) \cdot (D/\ell + k)$$
$$= \quad O((C + \min\{D, n\} \log n) \cdot (D/\min\{D, n\} + \lceil D/n \rceil))$$
$$\overset{C \leq n}{=} \quad O(C + D \log n) \ .$$

This completes the proof of Theorem 7.1.1.   □

7.1.2 Applications

The random delay protocol together with Corollary 5.2.2 yields the following result.

Corollary 7.1.1. *For any network G of size N with routing number R there exists an online protocol that routes any h-relation in time $O(R(h + \log n))$, w.h.p.*

This time bound is optimal for all networks of constant degree and $h \geq \log n$. However, if $h < \log n$ we obtain non-optimal results.

7.1.3 Limitations

The runtime bound of the random delay protocol holds for arbitrary, even non-simple, path collections. However, the definition of C must be changed for non-simple path collections in a way that if a packet traverses an edge e q times then it has to count q times for the congestion at e.

In the following we show how a multiplicative factor of $\log n$ in the time bound can be avoided when routing packets along more restricted classes of path collections. Let us start by demonstrating this for leveled path collections.

7.2 The Random Rank Protocol

The *random rank protocol* has its origin in a paper by Aleliunas [Al82] and Upfal [Up82], and can be found in a similar form as described below in Leighton's book [Le92]. It routes packets along an arbitrary leveled path collection of size n with congestion C and depth D in $O(C + D + \log n)$ steps, w.h.p., using edge buffers of size C. (Note that the depth of a leveled path collection is the number of levels formed by its nodes, and not necessarily its dilation.)

7.2.1 Description of the Protocol

At the beginning, every packet p gets a random rank denoted by rank(p) that is stored in its routing information. We require rank(p) to be chosen uniformly and independently from the choices of the other packets from some fixed range $[K]$ (K will be determined later). Additionally, each packet stores an identification number id(p) $\in [n]$ in its routing information that is different from all identification numbers of the other packets. The random rank protocol uses the following contention resolution rule.

> **Priority rule:**
> It two or more packets contend to use the same link at the same time then the one with minimal rank is chosen.

If two packets have the same rank then, in order to break ties, the one with the lowest id wins. The protocol then works as follows in each time step

> For each link with nonempty buffer, select a packet according to the priority rule and send it along that link.

For the random rank protocol the following time bound has been shown (see, e.g., [Le92]).

Theorem 7.2.1. *Suppose we are given a leveled path collection $\mathcal{P}$ of size n with congestion C and depth D. Let $K \geq 8C$. Then the random rank protocol needs at most $O(C + D + \log n)$ time steps to finish routing in $\mathcal{P}$, w.h.p., using edge buffers of size C.*

Proof. We prove this theorem in a similar way as done in [MW96]. Consider the runtime of the random rank protocol to be at least $T \geq D + s$. We want to show that it is very improbable that s is large. For this we need to find a structure that witnesses a large s. This structure should become more and more unlikely to exist the larger s becomes.

Let p_1 be a packet that arrived at its destination v_1 in step T. We follow the path of p_1 backwards until we reach a link e_1, where it was delayed the last time. Let us denote the length of the path from the destination of p_1 to e_1 (inclusive) by ℓ_1, and the packet that delayed p_1 by p_2. From e_1 we follow the path of p_2 backwards until we reach a link e_2 where p_2 was delayed the last time by some packet p_3. Let us denote the length of the path from e_1 (exclusive) to e_2 (inclusive) by ℓ_2. We repeat this construction until we arrive at a packet p_{s+1} that prevented the packet p_s at edge e_s from moving forward. Altogether it holds for all $i \in \{1, \ldots, s\}$: packet p_{i+1} leaves the buffer of e_i at time step $T - \sum_{j=1}^{i}(\ell_j + 1) + 1$, and prevents at that time step p_i from moving forward.

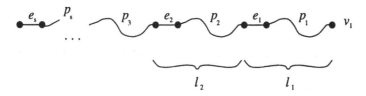

Fig. 7.1. The structure of a delay path.

The path from e_s to v_1 recorded by this process in reverse order is called *delay path* (see Figure 7.1). It consists of s contiguous parts of routing paths of length $\ell_1, \ldots, \ell_s \geq 0$ with $\sum_{i=1}^{s} \ell_i \leq D$. Because of the contention resolution rule it holds $\text{rank}(p_i) \geq \text{rank}(p_{i+1})$ for all $i \in \{1, \ldots, s\}$. A structure that contains all these features is defined as follows.

Definition 7.2.1 (s-delay sequence). *An s-delay sequence consists of*

- *s not necessarily different delay links $e_1, \ldots, e_s$;*
- *$s + 1$ delay packets $p_1, \ldots, p_{s+1}$ such that the path of p_i traverses e_i and e_{i-1} in that order for all $i \in \{2, \ldots, s\}$, the path of p_{s+1} contains e_s, and the path of p_1 contains e_1;*
- *s integers $l_1, \ldots, l_s \geq 0$ such that l_1 is the number of links on the path of p_1 from e_1 (inclusive) to its destination, for all $i \in \{2, \ldots, s\}$ l_i is the number of links on the path of p_i from e_i (inclusive) to e_{i-1} (exclusive), and $\sum_{i=1}^{s} l_i \leq D$; and*
- *$s + 1$ integers $r_1, \ldots, r_{s+1}$ with $0 \leq r_{s+1} \leq \ldots \leq r_1 < K$.*

A delay sequence is called active *if for all $i \in \{1, \ldots, s+1\}$ we have rank$(p_i) = r_i$.*

Our observations above yield the following lemma.

Lemma 7.2.1. *Any choice of the ranks that yields a routing time of $T \geq D + s$ steps implies an active s-delay sequence.*

Proof. Suppose the random rank protocol needs $T \geq D + s$ steps. Then we get for $\sum_{i=1}^{s} l_i \leq D$ that $T \geq \sum_{i=1}^{s} l_i + s$ and therefore $T - \sum_{i=1}^{s} l_i - s \geq 0$. Hence we can construct an active delay sequence of length s such that packet p_{s+1} leaves the buffer of e_s at time step $T - \sum_{i=1}^{s} (l_i + 1) + 1 \geq 1$. From this the lemma follows. □

Lemma 7.2.2. *The number of different s-delay sequences is at most*

$$n \cdot C^s \cdot \binom{D+s}{s} \cdot \binom{s+K}{s+1}.$$

Proof. There are at most $\binom{D+s}{s}$ possibilities to choose the l_i such that $\sum_{i=1}^{s} l_i \leq D$. Furthermore, there are n packets from which p_1 can be chosen. Since p_1 and l_1 determine the link e_1 and the congestion at e_1 is at most C, there are at most C possibilities to choose packet p_2. The same holds for the packets $p_3, \ldots, p_{s+1}$ at the edges $e_2, \ldots, e_s$. Hence we altogether have at most $\binom{D+s}{s} \cdot n \cdot C^s$ possibilities to choose the delay packets. Finally, there are at most $\binom{s+K}{s+1}$ ways to select the r_i such that $0 \leq r_{s+1} \leq \ldots \leq r_1 < K$. □

Note that during the execution of the random rank protocol the packets have a unique ordering w.r.t. their priority levels. (If two or more packets have the same rank, then the id's of the packets are compared.) Hence the packets in an s-delay sequence must be different. Since the packets choose their ranks independently at random, the probability that an s-delay sequence is active is $1/K^{s+1}$. Thus

Pr[The random rank protocol needs at least $D + s$ steps]

$$\overset{\text{Lemma 7.2.1}}{\leq} \quad \text{Pr[there exists an active } s\text{-delay sequence]}$$

$$\overset{\text{Lemma 7.2.2}}{\leq} \quad n \cdot C^s \cdot \binom{D+s}{s} \cdot \binom{s+K}{s+1} \cdot \frac{1}{K^{s+1}}$$

$$\leq \quad n \cdot C^s \cdot 2^{D+s} \cdot 2^{s+K} \cdot \frac{1}{K^{s+1}}$$

$$\leq \quad n \cdot 2^{2s+D+K} \cdot \left(\frac{C}{K}\right)^s$$

If we set $K \geq 8C$ and $s = K + D + (\alpha + 1)\log n$, where $\alpha > 0$ is an arbitrary constant, then

Pr[The random rank protocol needs at least $D + s$ steps]

$$\leq \quad n \cdot 2^{2s+D+K} \cdot 2^{-3s} = n \cdot 2^{-s+D+K} = \frac{1}{n^\alpha}$$

which concludes the proof of Theorem 7.2.1. □

7.2.2 Applications

The random rank protocol and Valiant's trick together yield the following result (see also [MV96]).

Theorem 7.2.2. *For any wrapped leveled path system of size N (that is, N input/output nodes) with expected congestion $\bar{C}$ and depth D there is an online protocol that routes any h-relation from the inputs to the outputs in time $O(h\bar{C} + D + \log N)$, w.h.p., using buffers of size $O(h\bar{C} + \log N)$.*

Proof. Consider first the problem of bounding the runtime of the random rank protocol for a randomly chosen function. For this we slightly have to modify the proof of Theorem 7.2.1. In addition to the probability that the packets chosen for a delay sequence have suitable ranks, we also have to consider the probability that the packets traverse the pieces of paths chosen for the delay path. This will be done in the following lemma which replaces Lemma 7.2.2.

Lemma 7.2.3. *For randomly chosen destinations, the expected number of different s-delay sequences is at most*

$$n \cdot \bar{C}^s \cdot \binom{D+s}{s} \cdot \binom{s+K}{s+1} \cdot$$

Proof. As before, there are at most $\binom{D+s}{s}$ possibilities to choose the ℓ_i such that $\sum_{i=1}^{s} \ell_i \leq D$, and at most $\binom{s+K}{s+1}$ ways to select the r_i such that $0 \leq r_{s+1} \leq \ldots \leq r_1 < K$. Furthermore, there are n possibilities to choose p_1. Let

e_1 be the link on the routing path of p_1 that has a distance of ℓ_1 from its destination. If p_1 is fixed then e_1 is fixed as well.

Now suppose that, for some $i \in \{1, \ldots, s\}$, p_j and e_j are fixed for all $1 \leq j \leq i$. The problem is to bound the expected number of candidates for p_{i+1}. Clearly, the routing path of p_{i+1} must traverse e_i, and p_{i+1} must be distinct from $p_1, \ldots, p_i$ because of the contention resolution rule. Let P be the set of all packets. Since each packet chooses a random destination, the expected number of possibilities to choose p_{i+1} is at most

$$\sum_{p_{i+1} \in P \setminus \{p_1, \ldots, p_i\}} \Pr[p_{i+1} \text{ traverses } e_i]$$

$$\leq \sum_{p_{i+1} \in P} \Pr[p_{i+1} \text{ traverses } e_i] = \bar{C}_{e_i} \leq \bar{C}$$

Let e_{i+1} be the link on the routing path of p_{i+1} that has a distance of ℓ_{i+1} from e_i. Once p_{i+1} is fixed then e_{i+1} is fixed as well.

Altogether, the expected number of choices for $p_2, \ldots, p_{s+1}$ (and therefore $e_2, \ldots, e_s$) is at most $\bar{C}^s$. Putting all pieces together yields the lemma. □

Since the packets choose their ranks independently at random, the probability that an s-delay sequence is active is $1/K^{s+1}$. Thus

$$\Pr[\text{The random rank protocol needs} \geq D + s \text{ steps}]$$

$$\leq \Pr[\text{there exists an active } s\text{-delay sequence}]$$

$$\leq n \cdot \bar{C}^s \cdot \binom{D+s}{s} \cdot \binom{s+K}{s+1} \cdot \frac{1}{K^{s+1}}$$

The rest is similar to the proof of Theorem 7.2.1.

As the expected congestion for routing h packets from each input node to random intermediate destinations, and from random intermediate destinations to their true destinations is both times bounded by $h\bar{C}$, the runtime bound of the theorem follows. Because the intermediate destinations of the packets are chosen independently at random, applying Chernoff bounds yields that the number of packets that want to traverse a link during the routing is bounded by $O(h\bar{C} + \log N)$, w.h.p. From this the bound for the buffer size stated in the theorem follows. □

Theorem 7.2.2 can be applied, for instance, to routing in butterflies.

Theorem 7.2.3. *Every h-relation can be routed in a wrap-around butterfly with N input/output nodes in time $O(h + \log N)$, w.h.p., using buffers of size $O(h + \log N)$.*

Proof. Let us consider the following path system for connecting the inputs with the outputs in a wrap-around butterfly: For each pair of nodes $((0, \alpha), (0, \beta))$, where $\alpha = (a_{d-1} \ldots a_1 a_0)_2$ and $\beta = (b_{d-1} \ldots b_1 b_0)_2$, the path from $(0, \alpha)$ to (d, β) contains the nodes $(0, a_{d-1} \ldots a_1 a_0)$, $(1, a_{d-1} \ldots a_1 b_0)$, $(2, a_{d-1} \ldots b_1 b_0) \ldots (d-1, a_{d-1} b_{d-2} \ldots b_0)$, $(0, b_{d-1} \ldots b_0)$ in this order. This path system has the following property.

Lemma 7.2.4. *Consider any wrap-around butterfly with a path system as described above. Then the expected congestion of routing a random h-function from the inputs to the outputs is $\frac{1}{2}h$.*

Proof. Consider an arbitrary edge e from level i to level $(i + 1)$ mod d of a d-dimensional wrap around butterfly, $i \in [d]$. There are altogether 2^i nodes in the top row that can reach e via paths in the path system described above. Since we consider random h-functions, each of the h packets stored in these nodes has a probability of $(\frac{1}{2})^{i+1}$ of choosing a path that runs through e. Altogether this yields an expected congestion of $\frac{1}{2}h$.   □

Using this result together with Theorem 7.2.2 yields Theorem 7.2.3.   □

This means that any global permutation can be routed in a wrap-around butterfly of size N in optimal time $O(\log N)$, w.h.p.

7.2.3 Limitations

The following observation shows that there are simple path systems for which the random rank protocol performs poorly. Its proof can be found in [LMR94].

Observation 7.2.1. *There exists a simple path collection of size n with dilation $D = O(\log n / \log \log n)$ and congestion $C = O(\log n / \log \log n)$, where the expected routing time of the random rank protocol is bounded by $\Omega((\log n / \log \log n)^{3/2})$.*

The path collection used for this observation consists of many subcollections of paths. Each subcollection consists of a linear array of length D, with loops of length $\sqrt{D}$ between adjacent nodes (see Figure 7.2).

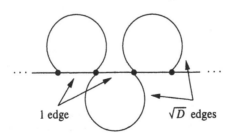

Fig. 7.2. The counterexample.

The packets traversing each subcollection of paths are broken into $\sqrt{D}$ groups numbered 0 through $\sqrt{D} - 1$ of $\sqrt{D}$ packets each. The packets in group i use the linear array for $i\sqrt{D}$ steps and then use $\sqrt{D}-i$ loops as their path. Note that if, for all $i \geq 0$, the packets in group i have smaller ranks than the packets in groups with larger numbers, than the packets in group i delay the packets in group $i+1$ by $D-(i+1)\sqrt{D}+i$ steps.

If we bound the buffer size then the following result can be shown.

Observation 7.2.2. *There exists a leveled path collection consisting of n paths with dilation $D = O(\log n/\log\log n)$ and congestion $C = O(\log^k n)$ for any constant $k > 0$ such that the expected runtime of the random rank protocol for buffers of size one is bounded by $\Omega(C \cdot D)$.*

Proof. For any fixed $k > 0$, let $c = \log^k n$ and $d = \frac{\log n}{4(k+1)\log\log n}$. Consider the path collection described in Figure 7.3 to be used by the sets of packets $P = \{p_1,\ldots,p_d\}$, $S = \{s_1,\ldots,s_d\}$, and $Q_i = \{q_1^i,\ldots,q_c^i\}$ for all $i \in \{1,\ldots,d\}$.

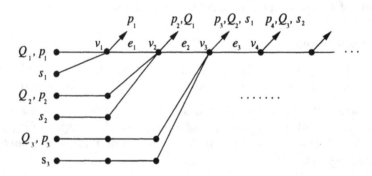

Fig. 7.3. The counterexample.

Clearly, this path collection has congestion $C = \Theta(c)$ and depth $D = \Theta(d)$. Let the id's of the packets be chosen such that all packets have different id's in $[n]$, and $p_1,\ldots,p_s$ have lower id's than packets in $Q_1,\ldots,Q_d$ and $s_1,\ldots,s_d$. Further, $\text{id}(s_d) < \text{id}(s_{d-1}) < \ldots < \text{id}(s_1)$. Let the id-rank of a packet p be defined as $\text{rank}(p) + \frac{\text{id}(p)}{n}$. Consider the following bad event E:

E: id-ranks of $p_1,\ldots,p_d$ < id-ranks of packets in $Q_1,\ldots,Q_d$ < id-rank(s_d) < $\ldots$ < id-rank(s_1).

Since according to the random rank protocol all packets start at the same time, the packets p_1 and s_1 arrive at the same time at node v_1. Because of the choice of the ranks, s_1 is preferred against p_1 to be forwarded along edge e_1. At node v_2, the packets p_2, s_2 and s_1 therefore arrive at the same time. Since id-rank(s_1) > id-rank(s_2), s_2 is preferred against s_1 to be forwarded along edge e_2. The same holds for the other pairs s_i, s_{i+1} for $i > 2$. Thus s_i prevents the packets in Q_i from moving forward for all $i \in \{1,\ldots,d\}$, and s_i

can only be moved forward along e_{i+1} if packet s_{i+1} and all packets in Q_{i+1} have been moved along link e_{i+1}. Hence the routing takes $\Omega(c \cdot d) = \Omega(C \cdot D)$ steps.

In the following we will bound the probability that such a bad event occurs. We start with computing a lower bound for the number of orderings of the $(c+2)d$ packets that the bad event. The number of ways to obtain such an ordering is $d!(c \cdot d)!$. Clearly, there are $((c+2)d)!$ possibilities to choose an arbitrary ordering for all $(c+2)d$ packets. For a randomly chosen order of id's and randomly chosen ranks, it holds because of symmetry reasons that every ordering of the id-ranks has the same probability to occur. If we choose the id's for the packets in such a way that they support the ordering in E, then we get a probability of at least

$$\frac{d!(c \cdot d)!}{((c+2)d)!} \geq \left(\frac{d}{((c+2)d)^2}\right)^d \left(\frac{c \cdot d}{(c+2)d}\right)^{c \cdot d} \geq \left(\frac{1}{d(c+2)^2}\right)^d \cdot e^{-2d}$$

that E is true. For $c = \log^k n$ and $d = \frac{\log n}{4(k+1)\log\log n}$ for any constant $k > 0$ this probability is at least $1/n^{2/3}$. Since a path collection of size n can consist of $\Theta(\frac{n}{c \cdot d})$ sub-collections as described in Figure 7.3, the expected number of sub-collections that fulfill E is at least one. This yields the observation.   □

The following protocol removes this drawback.

7.3 Ranade's Protocol

Consider the problem of routing packets in leveled networks with bounded degree and bounded buffers. As we will see, this can be done efficiently by a protocol we call *Ranade's protocol* (see also [Le92]). The difference between this and the previous protocol is that Ranade's protocol sends in addition to the usual packets so-called "ghost packets". This extension was first described by Ranade [Ra91] and generalized by Leighton *et al.* [LMRR94]. Ghost packets help the algorithm to maintain the following invariant:

A packet is routed along an edge only after all the other packets with lower ranks that must pass through the edge have done so.

7.3.1 Description of the Protocol

At the beginning every packet p gets a random rank$(p) \in [K]$ that is stored in its routing information (K is determined later). Furthermore a special end-of-stream (EOS) packet is inserted into the injection buffer, and is assigned rank K. Consider each *incoming* link to have space for storing at most $q \geq 1$ packets. In each time step, the algorithm then operates as follows.

For each node v with at least one nonempty buffer,

- select a packet p according to the priority rule (see Section 7.2.1),
- if the buffer of the next link on p's path contains fewer than q packets at the beginning of that step then send p out,
- send out a ghost packet with the rank of p along all other links,
- kill ghost packets that have been in a buffer in v at the beginning of that step.

Since each incoming link has its own buffer, a link buffer is therefore guaranteed never to hold more than q packets.

Ghost packets are sent to prevent buffers from getting empty. By complete induction it is easy to show that in this case all nodes send out packets in order of strictly increasing rank during the routing: If node v at level i sends out packets in order of strictly increasing rank, the rank of the ghost packets sent out along all other edges leaving v provides the receiving nodes at level $i+1$ with a lower bound on the ranks of the packets they will receive from v in the future. Therefore a node at level $i+1$ is also able to send out packets in order of strictly increasing rank.

End-of-stream packets are given a special treatment. Since an EOS packet has rank K, it can not be selected by a node unless there is only an EOS packet left in the node's injection buffer and in each of the buffers on all of the node's incoming links. Once an EOS packet has been selected, the node will create a new EOS packet for each of its outgoing edges, and each edge will attempt to send the corresponding packet at each step until it succeeds. After sending an EOS packet along an edge, a node will not send any more packets along that edge.

Ranade's protocol has the following runtime (see [LMRR94, Le92] for a proof).

Theorem 7.3.1. *Let G be an arbitrary network of constant degree. Suppose we are given a leveled path collection $\mathcal{P}$ of size n in G with congestion C, depth D, and edge buffers of any size $q \geq 1$. Let $K \geq \kappa \cdot C$ for some constant $\kappa \geq 1$. Then Ranade's protocol needs at most $O(C + D + \log n)$ time steps, w.h.p., to finish routing in $\mathcal{P}$.*

In case that the degree of G is not a constant then, according to [LMRR94], the runtime bound needs to be changed to $O(d \cdot 4^{(\log d)/q} C + D + \log n)$, w.h.p., where d denotes the maximum degree of G and q again the buffer size.

The proof of the runtime bound uses a delay sequence argument similar to the one used for bounding the runtime of the random rank protocol. However, in contrast to the random rank protocol there are two possible reasons why a packet may be delayed: It may be delayed either by a packet with lower rank staying in its current edge buffer, or by q packets with lower ranks occupying completely the next edge buffer on its way. That these q packets indeed have lower ranks is ensured by the ghost packets (see the invariant above). Hence we again can construct a sequence of packets with strictly decreasing ranks that witnesses a long runtime of the protocol. This monotonic behavior of the ranks can then be exploited to prove a good runtime bound. For more details see [LMRR94].

7.3.2 Applications

Similar to Theorem 7.2.2 the following result can be shown by using Valiant's trick.

Theorem 7.3.2. *For any wrapped leveled path system of size N (that is, N input/output nodes) with expected congestion $\bar{C}$ and depth D there is an online protocol that routes any h-relation in time $O(h\bar{C} + D + \log N)$, w.h.p., using buffers of size two.*

This theorem implies the following result together with Lemma 7.2.4.

Theorem 7.3.3. *Every h-relation can be routed in a wrap-around butterfly with N input nodes in time $O(h + \log N)$, w.h.p., using buffers of size two.*

Ranade [Ra91] also showed that any global permutation can be routed in a wrap-around butterfly of size N in optimal time $O(\log N)$ using only constant size buffers. This time bound can also be shown for permutation routing in shuffle-exchange networks [LMRR94]. Furthermore, Ranade's protocol can be used to route arbitrary d-relations in (n, d)-meshes in optimal time using only constant size edge buffers [LMRR94].

7.3.3 Limitations

As shown in [LMR94], Ranade's protocol has the following limitation.

Observation 7.3.1. *There is an N-node leveled network of degree 3 and a set of paths with congestion 3 and dilation 3 where the expected runtime of Ranade's protocol is $\Omega(\frac{\log^2 N}{\log \log N})$.*

The important point in the proof is that a path collection with congestion 3 and dilation 3 can have a large depth if the sources and destinations of the packets are not restricted to be at the same level. Hence Ranade's protocol can only be efficiently applied to path collections that have a dilation that is approximately the depth of the collection.

7.4 The Growing Rank Protocol

Now we present a protocol that routes packets along an arbitrary shortcut-free path collection of size n with congestion C and dilation D in $O(C + D + \log n)$ steps, w.h.p., using buffers of size C. It is called *growing rank protocol* [MV95] and works as follows.

7.4.1 Description of the Protocol

Initially, each packet is assigned an integer rank chosen randomly, independently, and uniformly from $[K]$. For each step, the protocol works as follows.

For each link with nonempty buffer,

- choose a packet p according to the priority rule,
- increase the rank of p by K/D, and
- move p forward along the link.

We generalize the result in [MV95] by analyzing the performance of the growing rank protocol for the following type of path collections.

Definition 7.4.1. *A path collection $\mathcal{P}$ is called d-shortcut-free if any piece of length at most d of a path in $\mathcal{P}$ can not be shortcut by any combination of other pieces of paths in $\mathcal{P}$.*

We can show the following theorem.

Theorem 7.4.1. *Suppose we are given a d-shortcut-free path collection $\mathcal{P}$ of size n with congestion C and dilation D, $d \leq D$. Let $K \geq 8C$. Then the growing rank protocol needs at most $O(C + \max\{1, \frac{\log(nD)}{d}\}D)$ time steps, w.h.p., to finish routing in $\mathcal{P}$.*

Proof. Similar to the proof of Theorem 7.2.1 we want to find a structure that witnesses a long runtime of the growing rank protocol. First we introduce the following definitions.

In the following, we denote the rank of a packet p while waiting to traverse a link e by $\text{rank}^e(p)$. Let $\text{id} : \{\text{set of packets}\} \to [n]$ be an arbitrary bijective function. We define the *ident-rank* of p at e as $\text{rank}^e(p) + \frac{\text{id}(p)}{n}$ and denote it by $\text{id-rank}^e(p)$. Note that in each round the ident-ranks of all packets are distinct. This type of rank ensures that whenever a packet p delays a packet p' at a link e it holds $\text{id-rank}^e(p) < \text{id-rank}^e(p')$. The following lemma shows that the rank of any packet can not be greater than $2K-1$ during the routing.

Lemma 7.4.1. *Suppose p is a packet which is stored in the buffer of link e in some round. Then $\text{rank}^e(p) \leq 2K - 1$.*

Proof. At the beginning, the rank of p is at most $K - 1$. Since the length of the routing path of p is at most D, the rank of p is increased by $\frac{K}{D}$ for at most D times. Thus, $\text{rank}^e(p) \leq K - 1 + D \cdot \frac{K}{D} \leq 2K - 1$. □

Analogous to the proof for the random rank protocol, the following delay sequence will serve as a witness for a long runtime of the growing rank protocol.

Definition 7.4.2 ((s, ℓ, K)-delay sequence). *An (s, ℓ, K)-delay sequence consists of*

1. s *not necessarily distinct* delay links $e_1, \ldots, e_s$;
2. $s + 1$ delay packets $p_1, p_2, \ldots, p_{s+1}$ *such that the path of p_i moves along the link e_i and the link e_{i-1} in that order for all $i \in \{2, \ldots, s\}$, the path of p_{s+1} contains e_s, and the path of p_1 contains e_1;*
3. s *integers $\ell_1, \ell_2, \ldots, \ell_s \geq 0$ such that ℓ_1 is the number of links on the routing path of packet p_1 from e_1 (inclusive) to its destination, for all $i \in \{2, \ldots, s\}$ ℓ_i is the number of links on the routing path of packet p_i from link e_i (inclusive) to link e_{i-1} (exclusive), and $\sum_{i=1}^{s} \ell_i \leq \ell$; and*
4. s *integer keys $r_1, r_2, \ldots, r_{s+1}$ such that $0 \leq r_{s+1} \leq \cdots \leq r_2 \leq r_1 < 2K$.*

We call s the length of the delay sequence. Further we say that a delay sequence is active, if $\mathrm{rank}^{e_i}(p_i) = r_i$ for all $i \in \{1, \ldots, s\}$ and $\mathrm{rank}^{e_s}(p_{s+1}) = r_{s+1}$

Lemma 7.4.2. *Suppose the routing takes $T \geq 2D$ or more rounds. Then there exists an active $(T - 2D, 2D, K)$-delay sequence.*

Proof. First, we give a construction scheme for a delay sequence. Let p_1 be a packet that arrived at its destination v_1 in step T. We follow p_1's routing path backwards to the last link on this path where it was delayed. We call this link e_1, and the length of the path from v to e_1 (inclusive) ℓ_1. Let p_2 be the packet that caused the delay, since it was preferred against p_1. We now follow the path of p_2 backwards until we reach a link e_2 at which p_2 was forced to wait, because the packet p_3 was preferred. Let us call the length of the path from e_1 (exclusive) to e_2 (inclusive) ℓ_2. We change the packet again and follow the path of p_3 backwards. We can continue this construction until we arrive at a packet p_{s+1} that prevented the packet p_s at edge e_s from moving forward.

The path from e_s to v_1 recorded by this process in reverse order is called *delay path*. It consists of contiguous parts of routing paths. In particular, the part of the delay path from link e_i (inclusive) to link e_{i-1} (exclusive) is a subpath of the routing path of packet p_i.

Let $r_i = \mathrm{rank}^{e_i}(p_i)$ for all $1 \leq i \leq s$ and $r_{s+1} = \mathrm{rank}^{e_s}(p_{s+1})$. Because of the contention resolution rule we have $0 \leq r_{s+1} \leq \cdots \leq r_1$, and $r_1 \leq 2K - 1$ because of Lemma 7.4.1. Thus, we have constructed an active (s, ℓ, K)-delay sequence for every $\ell \geq \sum_{i=1}^{s} \ell_i$.

Our next goal is to bound the sum of the ℓ_i's. In addition to the ranks $r_1, \ldots, r_{s+1}$, we denote by r_0 the rank of p_1 at its destination. It follows immediately from the protocol that $r_i + \ell_i \cdot \frac{K}{D} \leq r_{i-1}$ for all $1 \leq i \leq s$. As a consequence,

$$\sum_{i=1}^{s} \ell_i \cdot \frac{K}{D} \leq r_0 \overset{\text{Lemma 7.4.1}}{\Longrightarrow} \sum_{i=1}^{s} \ell_i \leq (2K - 1) \cdot \frac{D}{K} \leq 2D . \qquad (7.1)$$

Since the delay sequence consists of $\sum_{i=1}^{s} \ell_i$ moves and s delays, it covers at most $t = \sum_{i=1}^{s} \ell_i + s$ time steps. It follows that

$$t = \sum_{i=1}^{s} \ell_i + s \overset{(7.1)}{\leq} 2D + s \ .$$

Consequently, if we stop the above construction at packet p_{T-2D+1}, we still have $t \leq T$ and therefore found an active $(T-2D, 2D, K)$-delay sequence. □

Instead of considering the whole delay sequence, we will only consider a piece of it that is chosen in such a way that we can be sure that no packet can appear twice in it. For this we introduce the following definition.

Definition 7.4.3 $((s', \ell', K')$-delay subsequence). *An (s', ℓ', K')-delay subsequence consists of*

1. *s' not necessarily distinct delay links $e_1, \ldots, e_{s'}$;*
2. *$s' + 1$ delay packets $p_1, p_2, \ldots, p_{s'+1}$ such that the path of p_i moves along the link e_i and the link e_{i-1} in that order for all $i \in \{2, \ldots, s'\}$, the path of $p_{s'+1}$ contains $e_{s'}$, and the path of p_1 contains e_1;*
3. *s' integers $\ell_1, \ell_2, \ldots, \ell_{s'} \geq 0$ such that ℓ_i is the number of links on the routing path of packet p_i from link e_i (inclusive) to link e_{i-1} (exclusive) for all $i \in \{2, \ldots, s'\}$, and $\sum_{i=2}^{s'} \ell_i \leq \ell'$; and*
4. *s' integer keys $r_1, r_2, \ldots, r_{s'+1}$ such that $0 \leq r_{s'+1} \leq \cdots \leq r_2 \leq r_1 < r_{s'+1} + 2K'$ and $r_1 < 2K$.*

We say that a delay subsequence is active, *if $\mathrm{rank}^{e_i}(p_i) = r_i$ for all $i \in \{1, \ldots, s'\}$ and $\mathrm{rank}^{e_{s'}}(p_{s'+1}) = r_{s'+1}$*

The following lemma will be helpful to bound the total delay, length, and delay range of a subsequence of a delay sequence. Its proof is similar to a proof in [LMRR94] (see Lemma 2.10).

Lemma 7.4.3. *If there exists an active (s, ℓ, K)-delay sequence, then there exists an active $(\frac{s}{2\alpha}, \frac{2\ell}{\alpha}, \frac{2K}{\alpha})$-delay subsequence for every $\alpha \geq 1$.*

Proof. Suppose that an (s, ℓ, K)-delay sequence is active. Divide the packet sequence $p_2, \ldots, p_{s+1}$ into α contiguous subsequences such that each subsequence has at least $\lfloor s/\alpha \rfloor \geq s/2\alpha$ packets. This also partitions the delay path into subpaths. Let subsequence 0 consist only of packet p_1. For every subsequence $i \geq 1$, let ℓ_i denote the length of the ith subpath and let $2K_i$ denote the delay range of ranks for the ith subsequence, i.e., $2K_i$ is the difference between the rank of the last packet in subsequence $i-1$ when delayed by the first packet in subsequence i, and the rank of the last packet in subsequence i when delaying the second last. We know that there must be fewer than $\alpha/2$ segments with $K_i > 2K/\alpha$, since $\sum 2K_i \leq 2K$. Furthermore there must be fewer than $\alpha/2$ segments satisfying $\ell_i > 2\ell/\alpha$, since $\sum \ell_i \leq \ell$. Thus there must exist some segment for which $\ell_i \leq 2\ell/\alpha$ and $K_i \leq 2K/\alpha$. □

Next we show that, if we restrict $\frac{2K}{\alpha}$ to be at most $\frac{d}{2} \cdot \frac{D}{K}$, then no packet can appear twice in a $(\frac{s}{2\alpha}, \frac{2\ell}{\alpha}, \frac{2K}{\alpha})$-delay subsequence.

Lemma 7.4.4. *For any* (s', ℓ', K')-*delay subsequence with* $K' \leq \frac{d}{2} \cdot \frac{D}{K}$ *it holds that no packet can appear twice in it.*

Proof. Suppose, in contrast to our claim, that there is some packet p appearing twice in an (s', ℓ', K')-delay sequence. Then there exist i and j with $1 \leq i < j \leq s' + 1$ and $p = p_i = p_j$. Thus, the routing path of p crosses the delay path at the delay links e_{j-1} and e_i in that order. Since the rank of a packet is increased by $\frac{K}{D}$ each time it traverses an edge and the range of the ranks is bounded by $d \cdot \frac{D}{K}$, the length of the path the packet p traverses from e_{j-1} (inclusive) to e_i (exclusive) can be at most d. Let m denote the distance from link e_{j-1} (inclusive) to link e_i (exclusive) in the delay path. Since the routing paths are d-shortcut-free, the rank of p is increased at most m times while moving from e_{j-1} to e_i, and hence,

$$\text{id-rank}^{e_i}(p) \leq \text{id-rank}^{e_{j-1}}(p) + m \cdot \frac{K}{D} \ . \tag{7.2}$$

On the other hand, since for every $k \in \{1, \ldots, s'\}$ packet p_{k+1} delays packet p_k at edge e_k, we have id-rank$^{e_k}(p_k) >$ id-rank$^{e_k}(p_{k+1})$ for all $k \in \{1, \ldots, s'\}$. Further, the length of the routing path of packet p_{k+1} from e_{k+1} to e_k is ℓ_{k+1}, and thus the rank of p_{k+1} is increased by $\ell_{k+1} \cdot \frac{K}{D}$ on its path from e_{k+1} to e_k for all $k \in \{1, \ldots, s' - 1\}$. It follows that id-rank$^{e_k}(p_k) >$ id-rank$^{e_{k+1}}(p_{k+1}) + \ell_{k+1} \cdot \frac{K}{D}$ for all $k \in \{1, \ldots, s' - 1\}$. This yields

$$\text{id-rank}^{e_i}(p) > \text{id-rank}^{e_{j-1}}(p) + \sum_{k=i+1}^{j-1} \ell_k \cdot \frac{K}{D}$$

$$= \text{id-rank}^{e_{j-1}}(p) + m \cdot \frac{K}{D} \ . \tag{7.3}$$

Since (7.3) contradicts (7.2), there is no packet that appears twice in the delay subsequence. □

Our goal is therefore to restrict the range of the ranks used in the delay subsequence to be considered to at most $d \cdot \frac{D}{K}$. First we count the number of ways to construct an (s', ℓ', K')-delay subsequence.

Lemma 7.4.5. *The number of different* (s', ℓ', K')-*delay subsequences is at most*

$$n \cdot D \cdot 2K \cdot C^{s'} \cdot \binom{\ell' + s'}{s'} \cdot \binom{s' + 2K'}{s' + 1} \ .$$

Proof. There are n packets from which p_1 can be chosen, and at most D possibilities to choose ℓ_1. Furthermore there are at most $\binom{\ell'+s'}{s'}$ possibilities to choose the ℓ_i such that $\sum_{i=2}^{s} \ell_i \leq \ell'$. Since p_1 and ℓ_1 determine the link e_1 and the congestion at e_1 is at most C, there are at most C possibilities to choose packet p_2. The same holds for the packets $p_3, \ldots, p_{s'+1}$ at the edges $e_2, \ldots, e_{s'}$. Hence we altogether have at most $n \cdot D \cdot \binom{\ell'+s'}{s'} \cdot C^{s'}$ possibilities to choose the delay packets. Finally, there are at most $2K\binom{s'+2K'}{s'+1}$ ways to select the r_i such that $0 \le r_{s'+1} \le \ldots \le r_1 < r_{s'+1} + 2K'$ and $r_1 < 2K$. $\square$

Since the packets choose their ranks independently at random, the probability that an (ℓ', s', K')-delay subsequence is active is $1/K^{s'+1}$. Thus

$$\Pr[\text{there exists an active } (\ell', s', K')\text{-delay subsequence}]$$

$$\le \quad n \cdot D \cdot 2K \cdot C^{s'} \cdot \binom{\ell'+s'}{s'} \cdot \binom{s'+2K'}{s'+1} \cdot \frac{1}{K^{s'+1}}$$

$$\le \quad n \cdot D \cdot C^{s'} \cdot 2^{\ell'+s'} \cdot 2^{s'+2K'} \cdot \frac{1}{K^{s'}}$$

$$= \quad n \cdot D \cdot 2^{2s'+\ell'+2K'} \cdot \left(\frac{C}{K}\right)^{s'}$$

If we set $K \ge 8C$ and $s' \ge \ell' + 2K' + (\beta + 1)\log n + \log D$, where $\beta > 0$ is an arbitrary constant, then

$$\Pr[\text{there exists an active } (\ell', s', K')\text{-delay subsequence}]$$

$$\le \quad n \cdot D \cdot 2^{2s'+\ell'+2K'} \cdot 2^{-3s'} = n \cdot D \cdot 2^{-s'+\ell'+2K'} \le \frac{1}{n^\beta}$$

With $K' = \frac{d}{2} \cdot \frac{K}{D}$ we get from Lemma 7.4.3 that $\frac{d}{2} \cdot \frac{K}{D} = \frac{2K}{\alpha}$ and therefore $\alpha = \frac{4D}{d}$. Since any (s, ℓ, K)-delay sequence can have at most $2D$ edges, it holds that $\ell' \le \frac{4D}{\alpha} = d$, which has to be ensured for our analysis to work. Therefore the total delay s of the growing rank protocol is at most

$$2\alpha s' = 2\frac{4D}{d}\left(d + d \cdot \frac{K}{D} + (k+1)\log n + \log D\right)$$

$$= O\left(D + C + \frac{\log(nD)}{d}D\right) .$$

This concludes the proof of Theorem 7.4.1. $\square$

7.4.2 Applications

The proof of Theorem 7.4.1 can easily be modified to prove the following theorem by using Valiant's trick.

Theorem 7.4.2. *For any d-shortcut-free path system of size n with dilation D and expected congestion $\bar{C}$, $d \le D$, any permutation can be routed using the growing rank protocol in time $O(\bar{C} + \max\{1, \frac{\log(nD)}{d}\}D)$, w.h.p.*

Since any simple path system in a network with girth g is $(\lceil \frac{g}{2} \rceil - 1)$-shortcut-free (otherwise there must exist cycles of length smaller than g in it), we get the following result together with Theorem 5.1.1.

Theorem 7.4.3. *Let G be an arbitrary network of size N with routing number R and girth g. Then there exists a path system such that any h-relation can be routed in time $O((h + \max\{1, \frac{\log N}{g}\})R)$, w.h.p.*

Since the expander graphs constructed by Lubotzky *et al.* [LPS88] have a girth of $\Theta(\log N)$, this theorem implies that there exists an online protocol for these networks with runtime $O(\log N)$, w.h.p., for routing arbitrary permutations, which is optimal. Theorem 7.4.2 applied to node-symmetric networks yields the following result together with Theorem 5.1.2.

Theorem 7.4.4. *Let G be a node-symmetric network of size N with diameter D. Then the growing rank protocol routes packets according to an arbitrary h-relation in time $O(h \cdot D + \log N)$, w.h.p.*

Clearly, this time bound is optimal for permutation routing in arbitrary bounded degree node-symmetric networks. Together with Theorem 5.1.3 we get the following result for edge-symmetric networks.

Theorem 7.4.5. *Let G be an edge-symmetric network of size N with diameter D and degree d. Then the growing rank protocol routes packets according to an arbitrary h-relation in time $O((\frac{h}{d} + 1)D + \log N)$, w.h.p.*

7.4.3 Limitations

In case of bounded buffers, deadlocks can arise. Furthermore, the following observation can be shown (see [MV96]).

Observation 7.4.1. *Suppose C satisfies $\log n / \log \log n \leq C \leq n^\epsilon$ for some constant $\epsilon < 1$ and $C \geq D / \log \log n$. Then there is a simple path system of size n with dilation D and congestion C such that the expected routing time of the growing rank protocol is bounded by $\Omega(C + D \cdot \log n / \log \log n)$.*

The observation uses the following path collection.

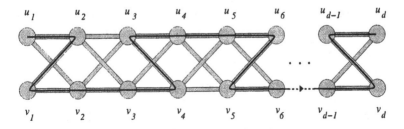

Fig. 7.4. The counterexample.

Let A and B be two sets of packets of size $C/2$ with source node u_1 and v_1 respectively. The routing path of the packets in $\mathcal{A}$ is

$$u_1 \to u_2 \to v_1 \to v_2 \to v_3 \to v_4 \to u_3 \to u_4 \to u_5 \to u_6 \to v_5 \to \dots$$
$$\dots \to v_{d-3} \to v_{d-2} \to v_{d-1} \to v_d \to u_{d-1} \to u_d$$

and the routing path of the packets in $\mathcal{B}$ is

$$v_1 \to v_2 \to u_1 \to u_2 \to u_3 \to u_4 \to v_3 \to v_4 \to v_5 \to v_6 \to u_5 \to \dots$$
$$\dots \to u_{d-3} \to u_{d-2} \to u_{d-1} \to u_d \to v_{d-1} \to v_d \ .$$

Since this path collection is at most 4-shortcut-free, the observation demonstrates that the analysis of the growing rank protocol above is nearly tight. Observation 7.4.1 further shows that the growing rank protocol can not be efficiently applied to arbitrary simple path collections. Note that it is still an open problem whether efficient shortcut-free path systems exist for any network. In case of shortest path systems, however, networks exist such that any shortest path system has a much higher expected congestion than the best simple path system (consider the union of a mesh and a complete binary tree). In the following we describe a protocol that can even route packets in optimal time along the path collection used for Observation 7.4.1.

7.5 The Extended Growing Rank Protocol

In this section we describe an extension of the growing rank protocol that can be efficiently used to route packets along the following type of path collections. It has been presented in [MS95a].

Let $\mathcal{P}$ be an arbitrary path collection. $\mathcal{P}$ is defined to have S *stages* of shortcut-free path collections, if stage numbers in $\{0, \dots, S-1\}$ can be assigned to the path collections in such a way that every path in $\mathcal{P}$ consists of subpaths that belong to path collections of strictly increasing stage number, and for all $s \in \{0, \dots, S-1\}$ the collection of all subpaths with stage number s forms a shortcut-free path collection. The *stage dilation* is defined as the length of the longest path in any of these S subpath collections, and the *stage congestion* is defined as the maximal congestion in any of these subpath collections. In the following we denote the stage congestion by C^* and the stage dilation by D^*.

7.5.1 Description of the Protocol

The extended growing rank protocol works as follows. Initially, each packet starting with a subpath that belongs to a shortcut-free path collection with stage number $q \in [S]$ is assigned an integer rank chosen uniformly at random from $\{q \cdot 2K, \dots, q \cdot 2K + (K-1)\}$, where K is determined later. In each time step, the protocol operates as follows.

For each link with nonempty buffer,

- choose a packet p according to the priority rule,
- if p changes from one shortcut-free path collection to another with stage number q', then choose a new random rank for p independently and uniformly at random from the set $\{q' \cdot 2K, \ldots, q' \cdot 2K + (K-1)\}$, otherwise increase the rank of p by K/D^*, and
- move p forward along the link.

This protocol yields the following result.

Theorem 7.5.1. *Suppose we are given an arbitrary path collection $\mathcal{P}$ of size n that can be partitioned into S shortcut-free path collections with stage congestion C^* and stage dilation D^*. Let $K \geq 8C^*$. Then the time the extended growing rank protocol needs to finish routing in $\mathcal{P}$ is at most $O(S(C^* + D^*) + \log n)$, w.h.p.*

Proof. Similar to the proof of Theorem 7.2.1 we want to find a structure that witnesses a long runtime of the extended growing rank protocol. First we introduce the following definitions.

In the following, we denote the rank of a packet p while waiting to traverse a link e by $\text{rank}^e(p)$. Let $\text{id} : \{\text{set of packets}\} \to [n]$ be an arbitrary bijective function. We define the *ident-rank* of p at e as $\text{rank}^e(p) + \frac{\text{id}(p)}{n}$ and denote it by $\text{id-rank}^e(p)$. Note that in each round the ident-ranks of all packets are distinct. This type of rank ensures that whenever a packet p delays a packet p' at a link e it holds $\text{id-rank}^e(p) < \text{id-rank}^e(p')$. The following lemma shows that the rank of any packet at stage q can not be greater than $2(q+1)K - 1$.

Lemma 7.5.1. *Suppose p is a packet at stage q which is stored in the buffer of link e in some round. Then $\text{rank}^e(p) \leq 2(q+1)K - 1$.*

Proof. At the beginning of stage q, the rank of p is at most $q \cdot 2K + K - 1$. Since the length of the routing path of p within one stage is at most D^*, the rank of p is increased by $\frac{K}{D^*}$ for at most D^* times. Thus, $\text{rank}^e(p) \leq q \cdot 2K + K - 1 + D^* \cdot \frac{K}{D^*} \leq 2(q+1)K - 1$. $\qquad\square$

Note that the rank of any packet during any stage of the routing is bounded above by $2S \cdot K - 1$. Analogous to the proof for the random rank protocol, the following delay sequence will serve as a witness for a long runtime of the extended growing rank protocol.

Definition 7.5.1 ((s, ℓ)-delay sequence). *An (s, ℓ)-delay sequence consists of*

1. *s not necessarily distinct delay links $e_1, \ldots, e_s$;*
2. *$s + 1$ delay packets $p_1, p_2, \ldots, p_{s+1}$ such that the path of p_i moves along the link e_i and the link e_{i-1} in that order for all $i \in \{2, \ldots, s\}$, the path of p_{s+1} contains e_s, and the path of p_1 contains e_1;*

3. s integers $\ell_1, \ell_2, \ldots, \ell_s \geq 0$ such that ℓ_1 is the number of links on the routing path of packet p_1 from link e_1 (inclusive) to its destination, for all $i \in \{2, \ldots, s\}$ ℓ_i is the number of links on the routing path of packet p_i from link e_i (inclusive) to link e_{i-1} (exclusive), and $\sum_{i=1}^{s} \ell_i \leq \ell$; and

4. s integer keys $r_1, r_2, \ldots, r_{s+1}$ such that $0 \leq r_{s+1} \leq \cdots \leq r_2 \leq r_1 < 2S \cdot K$.

We call s the length of the delay sequence. Further we say that a delay sequence is active, if $\mathrm{rank}^{e_i}(p_i) = r_i$ for all $i \in \{1, \ldots, s\}$ and $\mathrm{rank}^{e_s}(p_{s+1}) = r_{s+1}$.

Lemma 7.5.2. *Suppose the routing takes $T \geq 2S \cdot D^*$ or more rounds. Then there exists an active $(T - 2S \cdot D^*, 2S \cdot D^*)$-delay sequence.*

Proof. First, we give a construction scheme for a delay sequence. Let p_1 be a packet that arrived at its destination v_1 in step T. We follow p_1's routing path backwards to the last link on this path where it was delayed. We call this link e_1, and the length of the path from v_1 to e_1 (inclusive) ℓ_1. Let p_2 be the packet that caused the delay, since it was preferred against p_1. We now follow the path of p_2 backwards until we reach a link e_2 at which p_2 was forced to wait, because the packet p_3 was preferred. Let us call the length of the path from e_1 (exclusive) to e_2 (inclusive) ℓ_2. We change the packet again and follow the path of p_3 backwards. We continue this construction until we arrive at a packet p_{s+1} that prevented the packet p_s at edge e_s from moving forward.

The path from e_s to v_1 recorded by this process in reverse order is called *delay path*. It consists of contiguous parts of routing paths. In particular, the part of the delay path from link e_i (inclusive) to link e_{i-1} (exclusive) is a subpath of the routing path of packet p_i.

Let $r_i = \mathrm{rank}^{e_i}(p_i)$ for all $1 \leq i \leq s$ and $r_{s+1} = \mathrm{rank}^{e_s}(p_{s+1})$. Because of the contention resolution rule we have $0 \leq r_{s+1} \leq \ldots \leq r_1$, and $r_1 \leq 2S \cdot K - 1$ because of Lemma 7.5.1. Thus, we have constructed an active (s, ℓ)-delay sequence for every $\ell \geq \sum_{i=1}^{s} \ell_i$.

Our next goal is to bound the sum of the ℓ_i's. In addition to the ranks $r_1, \ldots, r_{s+1}$, we denote by r_0 the rank of p_1 at its destination. It follows immediately from the protocol that $r_i + \ell_i \cdot \frac{K}{D^*} \leq r_{i-1}$ for all $1 \leq i \leq s$. As a consequence,

$$\sum_{i=1}^{s} \ell_i \cdot \frac{K}{D^*} \leq r_0 \overset{\mathrm{Lemma\ 7.5.1}}{\Longrightarrow} \sum_{i=1}^{s} \ell_i \leq (2S \cdot K - 1) \cdot \frac{D^*}{K} \leq 2S \cdot D^* \ . \quad (7.4)$$

Since the delay sequence consists of $\sum_{i=1}^{s} \ell_i$ moves and s delays, it covers at most $t = \sum_{i=1}^{s} \ell_i + s$ time steps. It follows that

$$t = \sum_{i=1}^{s} \ell_i + s \overset{(7.4)}{\leq} 2SD^* + s \ .$$

Suppose that $T \geq 2SD^*$. If we stop the above construction at packet p_{T-2SD^*+1}, then we still have $t \leq T$ and therefore found an active $(T - 2SD^*, 2SD^*)$-delay sequence.   □

In order to bound the probability that a delay sequence is active, we need the following lemma.

Lemma 7.5.3. *If the routing paths of the packets are shortcut-free within any stage, then the tuples (p, q) of delay packets p at stage q in the above construction are pairwise distinct.*

Proof. Suppose, in contrast to our claim, that there is some packet p appearing twice at some stage q in the delay sequence. Then there exist i and j with $1 \leq i < j \leq s + 1$ and $p = p_i = p_j$. Thus, the routing path of p crosses the delay path at the delay links e_{j-1} and e_i in that order.

Let m denote the distance from link e_{j-1} (inclusive) to link e_i (exclusive) in the delay path. If the routing paths within each stage are shortcut-free, then the rank of p is increased at most m times while moving from e_{j-1} to e_i, and hence

$$\text{id-rank}^{e_i}(p) \leq \text{id-rank}^{e_{j-1}}(p) + m \cdot \frac{K}{D^*} . \tag{7.5}$$

On the other hand, each packet p_{k+1} delays the packet p_k at edge e_k, and consequently, $\text{id-rank}^{e_k}(p_k) > \text{id-rank}^{e_k}(p_{k+1})$ for all $k \in \{1, \ldots, s\}$. Further, the length of the routing path of packet p_{k+1} from e_{k+1} to e_k is ℓ_{k+1}, and thus the rank of p_{k+1} is increased by $\ell_{k+1} \cdot \frac{K}{D^*}$ on its path from e_{k+1} to e_k for all $k \in \{1, \ldots, s-1\}$. It follows that $\text{id-rank}^{e_k}(p_k) > \text{id-rank}^{e_{k+1}}(p_{k+1}) + \ell_{k+1} \cdot \frac{K}{D^*}$ for all $k \in \{1, \ldots, s-1\}$. This yields

$$\text{id-rank}^{e_i}(p) > \text{id-rank}^{e_{j-1}}(p) + \sum_{k=i+1}^{j-1} \ell_k \cdot \frac{K}{D^*}$$

$$= \text{id-rank}^{e_{j-1}}(p) + m \cdot \frac{K}{D^*} . \tag{7.6}$$

Since (7.6) contradicts (7.5), there is no packet that appears twice at some stage in the delay sequence.   □

Lemma 7.5.4. *The number of different (s, ℓ)-delay sequences is at most*

$$n \cdot (C^*)^s \binom{s+\ell}{s} \binom{s + 2S \cdot K}{s+1} .$$

Proof. We count the number of possible choices for each component:

- There are n possibilities to determine the packet p_1 and therefore the starting point v_1 of the delay path.
- Since $\sum_{i=1}^{s} \ell_i \leq \ell$, there are at most $\binom{s+\ell}{s}$ ways to choose the ℓ_i's.

- Furthermore, there are $\binom{s+2S \cdot K}{s+1}$ possibilities to choose the r_i's such that $0 \leq r_{s+1} \leq \ldots \leq r_1 < 2S \cdot K$.
- Once the ℓ_i's and r_i's are chosen, there are at most $(C^*)^s$ choices for the delay packets $p_2, \ldots, p_{s+1}$. This is because a fixed rank r_2 determines a fixed stage for p_2. Thus there are at most C^* choices for the packet p_2 given r_2. We follow the routing path of p_2 backwards for ℓ_2 steps, until we reach link e_2. Now we have at most C^* choices for p_3 given r_3. We follow again the routing path of this packet to link e_3, and so on, until we reach packet p_{s+1}.

So altogether, we find that the number of (s, ℓ)-delay sequences is at most

$$n \cdot (C^*)^s \binom{s+\ell}{s} \binom{s+2S \cdot K}{s+1} \ .$$

$\square$

Note that, according to Lemma 7.5.3, no packet appears twice at some stage in a delay sequence. Hence the ranks of the packets in any delay sequence have an offset in $[K]$ that is chosen independently at random. Therefore the probability that a particular delay sequence with $s + 1$ packets is active is at most $1/K^{s+1}$. As a consequence,

$$\Pr[\text{the routing takes } T = s + 2SD^* \text{ or more rounds}]$$

$$\overset{\text{Lemma 7.5.2}}{\leq} \Pr\left[\begin{array}{l} \text{an } (s, 2SD^*)\text{-delay sequence with distinct} \\ \text{delay packets within each stage is active} \end{array}\right]$$

$$\overset{\text{Lemma 7.5.4}}{\leq} n \cdot (C^*)^s \binom{s+\ell}{s} \binom{s+2S \cdot K}{s+1} \cdot \left(\frac{1}{K}\right)^{s+1}$$

$$\leq n \cdot 2^{s+\ell} \cdot 2^{s+2SK} \cdot \left(\frac{C^*}{K}\right)^s \ .$$

If we set $K \geq 8C^*$ and $s \geq \ell + 2SK + (\alpha + 1)\log n$, where $\alpha > 0$ is an arbitrary constant, then

$$\Pr[\text{the routing takes at least } T = s + 2SD^* \text{ steps}]$$

$$\leq \ n \cdot 2^{2s+\ell+2SK} \cdot 2^{-3s} \leq \frac{1}{n^\alpha} \ .$$

Since $\ell \leq 2SD^*$, this proves Theorem 7.5.1.

$\square$

7.5.2 Applications

In the following we show that the extended growing rank protocol can be used for efficient network emulations, which has applications in the field of compact routing.

Consider the problem of simulating the routing of an arbitrary permutation in a network G by a network H. Using Valiant's trick of first routing packets to random intermediate destinations in G before they are routed to their true destinations, it suffices to consider the problem of routing a random function in G. Let H be an arbitrary node-symmetric network of size N with diameter D_H. Further let G be an arbitrary network of size N with degree δ and a path system with dilation D_G and expected congestion γD_G.

We start with describing how to embed G in H. Consider an arbitrary 1–1 embedding of the nodes in G into the nodes of H. We intend to simulate every edge in G by a path in H that connects the nodes simulating its endpoints. Since the degree of G is δ, there is a path collection in H consisting of two stages of shortest path collections for simulating the edges in G with congestion at most $O(\delta D_H + \log N)$ and dilation at most D_H (use Theorem 5.1.2 together with Valiant's trick).

Consider now the problem of routing a random function in G. Suppose that each packet p chooses a random *startup stage* $s \in [D_G]$. Then we define a packet p to be at *superstage* $q \geq s$ at edge e if e is the $(q - s)$th edge on its path. Since $\mathcal{P}_G$ has dilation D_G, the number of different superstages used by the packets is at most $2D_G$. As the expected congestion of $\mathcal{P}_G$ is γD_G, the expected number of packets that visit some fixed edge e at some fixed superstage q is at most γ.

As noted above, each superstage can be simulated by H by moving the packets along two stages of shortcut-free paths. Thus altogether, simulating the routing of a random function in G by H using random startup stages in $[D_G]$ requires at most $4D_G$ stages of shortcut-free path collections. Since the stage dilation is at most D_H, it remains to calculate an upper bound for the stage congestion that holds w.h.p.

Consider some fixed superstage q. Since the expected number of packets that want to use some fixed edge e in G at some fixed superstage is γ, the expected number of packets that want to traverse a path in H at some fixed superstage is at most γ. As the congestion of the shortcut-free path collections in H simulating edges in G is at most $O(\delta D + \log N)$, the expected number of packets that want to use some fixed edge e in H at some fixed superstage is at most $\bar{C}^* = O(\gamma(\delta D_H + \log N))$.

For any fixed q, let the random variable X_e^q denote the number of packets at stage q that traverse e during the simulation of routing in G by H. Further let for every node i in H the binary random variable $X_{i,e}^q$ be one if and only if the packet with source i traverses e at stage q during the simulation. Then $X_e^q = \sum_{i \in [N]} X_{i,e}^q$. Since the probabilities for the $X_{i,e}^q$ are independent and $E(X_e^q) \leq \bar{C}^* = O(\gamma(\delta D_H + \log N))$, we can apply Chernoff bounds to get that the stage congestion for e is bounded by $O(\gamma(\delta D_H + \log N) + \log N)$, w.h.p. Thus the stage congestion for the whole simulation of routing a random function in G is bounded by $O(\gamma(\delta D_H + \log N) + \log N)$, w.h.p. Hence we get the following theorem.

Theorem 7.5.2. *Any node-symmetric network H of size N with diameter $D_H = \Omega(\log N)$ can simulate the routing of any permutation (or randomly chosen function) in a network G with degree δ and a path system with dilation D_G and expected congestion γD_G in*

$$O((\gamma\delta + 1)D_G \cdot D_H)$$

steps, w.h.p.

This result implies the following corollary.

Corollary 7.5.1. *Any node-symmetric network H of size N with diameter $D = \Omega(\log N)$ can simulate the routing of a randomly chosen function in any bounded degree network G of size N with routing number R in time $O(R \cdot D)$, w.h.p.*

The advantage of the simulation strategy used for this corollary is that it does not perform a step-by-step simulation. Note that in order to get the same time bound with a step-by-step simulation we would require a protocol for G that routes a randomly chosen function in time $O(R)$, w.h.p. For arbitrary bounded degree networks, however, such a protocol is not known yet.

Theorem 7.5.2 also has interesting consequences for simulating routing in s-ary butterflies by node-symmetric networks, since s-ary butterflies have routing strategies with a low expected congestion.

Lemma 7.5.5. *There is a randomized online strategy for the s-ary wrap-around butterfly of size N using paths of dilation at most $2\log_s N$ such that, for a randomly chosen function, the expected congestion at*

- *any node is at most $2\log_s N$, and*
- *any edge is at most $\frac{2\log_s N}{s}$.*

Proof. Consider an arbitrary s-ary d-dimensional wrap-around butterfly. For any packet p with source (ℓ, x) and destination (ℓ', y), we use the following strategy:

We first route p to a random intermediate destination at level (ℓ', z). This will be done by selecting at random one out of s possible edges to the next higher level until we reach level ℓ'. In order to get from (ℓ', z) to (ℓ', y), p uses the unique path of length d along the levels $\ell', \ell' + 1, \ell' + 2, \ldots, \ell'$. For a randomly chosen destination, this path corresponds to randomly choosing one out of s possible edges to the next higher level for each of the d levels.

Hence routing p to a randomly chosen destination as described above implies routing p along one out of s possible edges to the next higher level, chosen independently at random, for at most $2d$ levels. Since the s-ary wrap-around butterfly is node-symmetric, the expected congestion at every node is the same for a randomly chosen function, and therefore at most $2d$. Hence the expected congestion at every edge is at most $\frac{2d}{s}$. Since the dimension of any s-ary butterfly of size N is at most $\log_s N$, the lemma follows. $\qquad\square$

Lemma 7.5.5 and Theorem 7.5.2 together yield the following theorem.

Theorem 7.5.3. *Any node-symmetric network of size N with diameter $D = \Omega(\log N)$ can simulate any permutation routing problem on an s-ary wrap-around butterfly of size N in time $O(\log_s N \cdot D)$, w.h.p.*

Because of the regular structure of an s-ary butterfly the nodes only need to know the numbers of the subbutterflies their edges are connected with in order to send packets to their correct destinations. Since less space is necessary to store the paths in H simulating edges of the s-ary butterfly than to store complete routing tables in H, this strategy can be used to develop fast compact routing schemes for any node-symmetric network H. For a more detailed description of how this works we refer to the chapter on compact routing schemes.

7.5.3 Limitations

In contrast to the previously presented protocols, in order to apply Theorem 7.5.1 to a network we can not use a bound on the expected stage congestion for C^*, but instead need a bound for the stage congestion that holds w.h.p. The reason for this is that for the extended growing rank protocol it can happen, in contrast to the previous protocols, that the same packet appears more than once in an active delay sequence. Therefore, a high congestion at one edge may also imply a high congestion at another edge.

Until now it is not known whether there is a simple path collection for which the extended growing rank protocol performs poorly. However, the protocol still requires that parts of a path collection are shortcut-free. This drawback will be removed in the following protocol.

7.6 The Trial-and-Failure Protocol

In this section we present a protocol that routes packets along an arbitrary simple path collection of size n with link bandwidth B, congestion C, and dilation D in time

$$O\left(\frac{C \cdot D^{1/B} + D \log n}{B} + D\right)$$

w.h.p., without buffering. It is called *trial-and-failure protocol* and has been presented in [CMSV96]. Since it can be used to route along arbitrary simple path collections, we will apply the protocol to routing in arbitrary networks, using the routing number of a network to bound its runtime.

The idea of the trial-and-failure protocol is that, once a packet leaves its source, it has to move along the edges of its path without waiting until it reaches its destination. If too many packets want to use the same link at the same time then some are discarded (and therefore have to be rerouted). In the following we assume that every link has bandwidth $B \geq 1$.

7.6.1 Description of the Protocol

Consider an arbitrary simple path collection of size n with dilation D and link bandwidth B. Initially, each packet p_i chooses uniformly and independently from the other packets a random rank $r_i \in [K]$ (K will be specified later) and a random delay $d_i \in [D]$. Additionally, p_i stores an identification number $\text{id}(p_i) \in [n]$ in its routing information that is different from all identification numbers of the other packets. Let us define the following contention resolution rule.

B-priority rule:
If more than B packets attempt to use the same link during the same time step, then those B with lowest rank win.

If two or more packets have the same rank then, in order to break ties, the ones with the lowest id's win. Then the protocol works as follows.

repeat

- **forward pass**: Each active packet p_i waits for d_i steps. Then it is routed along its path, obeying the B-priority rule.
- **backward pass**: For each packet that reached its destination during the forward pass, an acknowledgment is sent back to the source. Upon receipt of the acknowledgment, the source declares the packet inactive.

until no packet is active

Clearly, the forward pass needs $2D$ steps to be sure that every packet that has not been discarded during the routing reaches its destination. These packets are called *successful*. In the backward pass, the forward pass is run in reverse order. Therefore, no collisions can occur in the backward pass, and $2D$ steps suffice to send all acknowledgments back.

Theorem 7.6.1. *Suppose we are given an arbitrary simple path collection $\mathcal{P}$ of size n with link bandwidth B, congestion C, and dilation D. Further let $K \geq 4C \cdot D^{-1+1/B}$. Then the trial-and-failure protocol needs*

$$O\left(\frac{C \cdot D^{1/B} + D \log n}{B} + D\right)$$

steps, w.h.p., to finish routing in $\mathcal{P}$ without using buffers.

Proof. In order to prove the time bound, we again use a delay sequence argument.

Definition 7.6.1 ((s, B)-delay sequence). *An (s, B)-delay sequence consists of*

1. $s \cdot B + 1$ *delay packets* $p_1, \ldots, p_{s \cdot B + 1}$ *such that during the forward pass of round* $i \geq 1$, *packet* $p_{i \cdot B + 1}$ *is prevented from using a link by packets* $p_{(i-1)B+1}, \ldots, p_{i \cdot B}$;
2. $s \cdot B + 1$ *integer keys* $r_1, \ldots, r_{s \cdot B + 1}$ *such that* $0 \leq r_1 \leq \ldots \leq r_{s \cdot B + 1} < K$.

A *delay sequence* is called active if $\mathrm{rank}(p_i) = r_i$ for all $1 \leq i \leq s \cdot B + 1$.

Lemma 7.6.1. *If the routing takes more than s rounds, there exists an active (s, B)-delay sequence.*

Proof. In the following we will give a construction for such a delay sequence. If the routing takes more than s rounds then there is a packet $p_{s \cdot B + 1}$ that did not reach its destination in round s. In such a case there must have been packets $p_{(s-1)B+1}, \ldots, p_{s \cdot B}$ that prevented $p_{s \cdot B + 1}$ from using some link in round s. According to the routing protocol, $p_{(s-1)B+1}, \ldots, p_{s \cdot B}$ can be chosen such that $\mathrm{rank}(p_{(s-1)B+1}) \leq \ldots \leq \mathrm{rank}(p_{s \cdot B}) \leq \mathrm{rank}(p_{s \cdot B + 1})$. But if $p_{(s-1)B+1}$ was still active in round s, there must have been packets $p_{(s-2)B+1}, \ldots, p_{(s-1)B}$ in round $s - 1$ that prevented $p_{(s-1)B+1}$ from using some link, where $\mathrm{rank}(p_{(s-2)B+1}) \leq \ldots \leq \mathrm{rank}(p_{(s-1)B}) \leq \mathrm{rank}(p_{(s-1)B+1})$. Continuing this argument back to round 1 yields the lemma with $r_i :=$ $\mathrm{rank}(p_i)$ for all $1 \leq i \leq s \cdot B + 1$. □

Lemma 7.6.2. *The number of different (s, B)-delay sequences is at most*

$$n \cdot (D \cdot C^B)^s \binom{s \cdot B + K}{s \cdot B + 1}$$

Proof. We count the number of possible choices for the components:

- There are at most $n \cdot (D \cdot C^B)^s$ choices for the delay packets. This is because there are at most n possibilities to determine the packet $p_{s \cdot B + 1}$, and if $p_{i \cdot B + 1}$ is fixed, there are at most $D \cdot C^B$ choices for the packets $p_{(i-1)B+1}, \ldots, p_{i \cdot B}$ for all $1 \leq i \leq s$.
- There are $\binom{(s \cdot B + 1) + K - 1}{s \cdot B + 1}$ possibilities to choose the r_i's such that $0 \leq r_1 \leq \ldots \leq r_{s \cdot B + 1} < K$.

Multiplying these terms yields the lemma. □

Since each packet chooses a random rank and a random delay, and the contention resolution rule of the above routing protocol requires that in any active delay sequence the packets have to be pairwise distinct, the probability that a particular (s, B)-delay sequence is active is at most $K^{-(s \cdot B + 1)}(1/D)^{s \cdot B}$. Therefore, if we choose

$$K \geq \frac{4C \cdot D^{1/B}}{D} \quad \text{and} \quad T \geq \frac{K + (\alpha + 1) \log n}{B} + 1$$

for some arbitrary constant $\alpha > 0$ then

Pr[the routing takes more than T rounds]

$$\overset{\text{Lemma 7.6.1}}{\leq} \quad \Pr\left[\text{a } (T, B)\text{-delay sequence is active}\right]$$

$$\overset{\text{Lemma 7.6.2}}{\leq} \quad n \cdot (D \cdot C^B)^T \binom{T \cdot B + K}{T \cdot B + 1} K^{-(T \cdot B + 1)} \left(\frac{1}{D}\right)^{T \cdot B}$$

$$\leq \quad n \cdot 2^{T \cdot B + K} \cdot \left(\frac{C \cdot D^{1/B}}{D \cdot K}\right)^{T \cdot B}$$

$$\leq \quad n \cdot 2^{T \cdot B + K} \cdot 2^{-2T \cdot B} \leq n \cdot 2^{-(\alpha+1)\log n} = \frac{1}{n^\alpha} \ .$$

This yields Theorem 7.6.1. □

Since $\Omega(\frac{C}{B} + D)$ is a lower bound for any protocol that routes packets along a simple path collection $\mathcal{P}$ of size n with link bandwidth B, congestion C, and dilation D, the runtime of the trial-and-failure protocol gets optimal for $C \geq D \log n$ and $B \geq \log D$, or $B \geq \log(nD)$. Note that in case of $C = O(D)$ and $B \geq \log D$ the range from which the ranks have to be chosen reduces to a constant. In particular, if $C \cdot D^{-1+1/B} \leq 1/4$ then the above protocol does not require random ranks any more. That is, only random delays are needed in this case to get the time bound in Theorem 7.6.1.

7.6.2 Applications

If we simulate link bandwidth by buffers we arrive at the following result.

Corollary 7.6.1. *Given any simple path collection of size n with buffer size B, congestion C, and dilation D, the trial-and-failure protocol requires $O(C \cdot D^{1/B} + D \log n)$ time to route all packets, w.h.p.*

This result is optimal if $C \geq D \log n$ and $B \geq \log D$. Together with Valiant's trick we get the following result.

Corollary 7.6.2. *Let G be an arbitrary network of size N with buffer size B and routing number R. Then the trial-and-failure protocol routes packets according to an arbitrary h-relation in time $O((h \cdot R^{1/B} + \log N)R)$, w.h.p.*

In case of shortcut-free path collections, Theorem 7.6.2 provides the following tradeoff between routing time and buffer size for oblivious routing.

Theorem 7.6.2. *Given any shortcut-free path collection of size n with congestion C, dilation D, and buffer size $B \leq C$ there exists an online routing protocol that requires*

$$O\left(\frac{C \cdot D^{1/B} + \log n}{B} \cdot (C + D + \log n)\right)$$

time to route all packets, w.h.p.

Proof. Consider routing n packets along a shortcut-free path collection with congestion C and dilation D in a network with buffer size B. Each packet p_i gets two random ranks: the first rank r_i^t is chosen as in the trial-and-failure protocol, and the second rank r_i^g is chosen as in the growing rank protocol. The new protocol then works as follows

> for each packet p_i, choose random ranks r_i^t and r_i^g, all n packets are declared active
> **repeat**
>
> - **forward pass:** route packets according to the growing rank protocol; in case of overfull buffers use the B-priority rule (that is, those B packets with lowest ranks r_i^t survive, the others are eliminated)
> - **backward pass:** successful packets become inactive via acknowledgments by reversing the forward pass
>
> **until** no packet is active

From Theorem 7.4.1 we know that the growing rank protocol requires at most $O(C + D + \log n)$ time, w.h.p., to be sure that packets either reached their destination or have been discarded because of an overfull buffer. Using a similar delay sequence argument for eliminations of packets as in the proof of Theorem 7.6.1 yields that $O(\frac{1}{B}(C \cdot D^{1/B} + \log n))$ rounds suffice to route all packets, w.h.p. (leave out the probability $\frac{1}{D}$ for the delay). This completes the proof of Theorem 7.6.2.   □

7.6.3 Limitations

The runtime bound of the trial-and-failure protocol holds for arbitrary, even non-simple, path collections. However, the definition of C must be changed for non-simple path collections in a way that if a packet traverses an edge e q times then it has to count q times for C_e.

7.7 The Duplication Protocol

In this section we present an efficient protocol for routing packets along an arbitrary simple path collection of size n with congestion C, dilation D, and bandwidth $\Theta(\log(C \cdot D)/\log\log(C \cdot D))$ without buffering. It is called *duplication protocol* and has been presented in [CMSV96]. Since it can be used to route along arbitrary simple path collections, we will apply the protocol to routing in arbitrary networks, using the routing number of a network to bound its runtime.

The idea of the duplication protocol is similar to that of the trial-and-failure protocol, with the difference that each new trial the number of copies sent out for a packet is duplicated. This significantly increases the chance that finally one of the copies will be able to reach the destination. In the following we again assume that every link has bandwidth $B \geq 1$.

7.7.1 Description of the Protocol

We define the following rule in case of contention between packets:

> **B-collision rule:**
> If more than B packets attempt to use the same link during the same time step, then *all* of them are discarded.

The duplication protocol consists of $k + 1$ rounds (the parameters k, C_r, and B will be determined later).

> **Duplication Protocol:**
> all n packets are declared active
> **for** $r = 0$ **to** k **do**
>
> - **forward pass:** for each active packet, send 2^r copies of it with random startup delays in $[C_r]$, route the copies according to the contention resolution rule above
> - **backward pass:** successful packets become inactive via acknowledgments

Clearly, the forward pass of round r requires $C_r + D$ steps to be sure that either a packet reaches its destination or has been discarded. The backward pass can be organized in such a way that acknowledgments of successful copies are not delayed or discarded by running the forward pass in reverse order. A packet becomes inactive if it receives an acknowledgment of at least one of its copies.

Theorem 7.7.1. *Given any simple path collection $\mathcal{P}$ of size n with congestion C, dilation D, and bandwidth $\Theta(\log(C \cdot D)/\log\log(C \cdot D))$, the duplication protocol requires*

$$O\left(C + D \log\log n + \frac{\log n \cdot \log\log n}{\log\log(C \cdot D)} \right)$$

time, w.h.p., to finish routing in $\mathcal{P}$ without buffering.

Proof. Our line of proof will be to show that the congestion is reduced enough from round to round such that more and more copies of each active packet can be sent out, until the probability that none of its copies reaches its destination gets polynomially small in the number of packets.

We first need the following definitions. A *site* is an ordered pair (e, s) where e is an edge and s is a step in the forward pass (of the round currently being considered). A packet p *aims for* a site (e, s) if p selects a random delay such that it will be present in edge e at step s of the forward pass, provided that it is not discarded before reaching edge e. Note that whether or not a packet aims for a site is strictly a function of its starting time and its path, and is independent of the starting times of other packets. A packet p *is discarded at a site* (e, s) if p attempts to enter edge e at step s but is discarded because of edge contention. Note that if y packets are discarded at a site (e, s), then at least $y > B$ packets aim for site (e, s). Given any packet p, p's *conflicting packets* consist of all packets with paths that intersect p's path (including p itself). A packet's *region* consists of all sites its conflicting packets could aim for.

We will begin by analyzing the forward pass of round 0. Let $C_0 = C$ and $B = 10\ln(C \cdot D)/\ln\ln(C \cdot D)$. Consider any packet p and mark any m sites in p's region. Let the random variable X denote the number of distinct packets that aim for any of the marked sites. Note that the expected number of packets which aim for any one site is at most one (because at most C packets can have the possibility of aiming for the site, and the probability that such a packet does aim for the site is $1/C$). Therefore, $E[X] \leq m$.

Because X is the sum of independent Bernoulli trials, we can use Chernoff bounds to bound the probability that X is larger than $B \cdot m$:

$$\Pr[X > B \cdot m] \leq (e/B)^{B \cdot m} = e^{-m \cdot B(\ln B - 1)}.$$

Since packet p has at most $C \cdot D$ conflicting packets, and for each of these packets there are at most $D(C + D)$ sites (e, s) it could visit, there are at most

$$\binom{C \cdot D^2(C + D)}{m} \leq \left(\frac{eC \cdot D^2(C + D)}{m}\right)^m \leq e^{m(3\ln(C \cdot D) + 1)}$$

different ways of marking m sites in p's region. Therefore, the probability that more than $B \cdot m$ distinct packets aim for *any* set of m sites in p's region is at most

$$e^{m(3\ln(C \cdot D) + 1 - B(\ln B - 1))}.$$

Let $m = (\alpha + 2)\ln n/2\ln(C \cdot D)$. Because

$$
\begin{aligned}
\ln B - 1 &= \ln(10/e) + \ln\ln(C \cdot D) - \ln\ln\ln(C \cdot D) \\
&\geq (1/2)\ln\ln(C \cdot D) + 1 ,
\end{aligned}
$$

we have

$$e^{m(3\ln(C \cdot D) + 1 - B(\ln B - 1))} \leq n^{-(\alpha+2)}.$$

Note that if more than $B \cdot m$ packets are discarded at sites in p's region, then more than $B \cdot m$ distinct packets must aim for some set of m sites in p's region. (This can be argued as follows: Let z be the number of sites in

p's region at which packets were discarded. If $z \leq m$, let S be a set of any m sites including these z sites. If $z > m$, let S be a set of any m of these z sites. Note that more than B packets were discarded at each of them. Therefore $|S| = m$, more than $B \cdot m$ packets were discarded at sites in S, each of these discarded packets aimed for the site at which it was discarded, and these discarded packets must be distinct because a packet can only be discarded once.) As a result, the probability that more than $B \cdot m$ packets are discarded at sites in *any* packet's region is at most $n^{-(\alpha+1)}$. We will say that round 0 succeeded iff for every packet p, at most $B \cdot m = 5(\alpha + 2) \ln n / \ln \ln(C \cdot D)$ of p's conflicting packets were not successfully delivered to their destinations.

Now consider round 1. It is identical to round 0 in terms of the number of (copies of) packets that are discarded, except that it has congestion at most $2B \cdot m$ (assuming that round 0 succeeded) because two copies are made of each of the at most $B \cdot m$ packets that were discarded at sites in each packet's region, and because the paths are simple, each of these packet copies can contribute at most one to the congestion of an edge. Therefore, we will set $C_1 = 2B \cdot m$ and an analysis identical to the analysis of round 0 shows that the probability that more than $B \cdot m$ packet copies are discarded at sites in *any* packet's region during round 1 is at most $n^{-(\alpha+1)}$. Because both copies of a packet must be discarded in order for the packet to fail to be delivered, it follows that the probability that there exists a packet p such that more than $B \cdot m/2$ of p's conflicting packets were not successfully delivered after round 1 is at most $n^{-(\alpha+1)}$.

In general, for any round r, $1 \leq r \leq k$, 2^r copies are made of each active packet, which results in a copy congestion of at most $C_r = 2B \cdot m$. We will say that round r succeeded iff for every packet p, at most $B \cdot m/2^r$ of p's conflicting packets were not successfully delivered to their destinations during the first r rounds. Using an analysis identical to that given for round 1, it follows that for any round r, $1 \leq r \leq k$, if all rounds $i < r$ succeeded then the probability that round r fails is at most $n^{-(\alpha+1)}$. Setting $k = \log(B \cdot m) + 1 = O(\log \log n)$, it follows that if all $k + 1$ rounds $0 \ldots k$ succeed, then every packet has been successfully delivered, since at round k every active packet gets $2B \cdot m$ copies. The probability that all $k + 1$ rounds succeed at least $(1 - 1/n^{\alpha+1})^{k+1} \geq 1 - (k + 1)/n^{\alpha+1} \geq 1 - n^{-\alpha}$.

Finally, note that the algorithm requires link bandwidth $B = \Theta(\log(C \cdot D)/\log \log(C \cdot D))$ and takes $2(C + D) + 2k(C_1 + D) = O(D \log \log n + C + \log n \log \log n / \log \log(C \cdot D))$ time, w.h.p. This completes the proof of Theorem 7.7.1. □

7.7.2 Applications

If we simulate bandwidth B by buffer size B, we arrive at the following result.

Corollary 7.7.1. *Given any simple path collection $\mathcal{P}$ of size n with congestion C and dilation D, the duplication protocol can be used to finish routing* *in* $\mathcal{P}$ *in*

$$O\left(\left(C + D\log\log n + \frac{\log n \cdot \log\log n}{\log\log(C \cdot D)}\right) \cdot \frac{\log(C \cdot D)}{\log\log(C \cdot D)}\right)$$

time, w.h.p., using buffers of size $\Theta(\log(C \cdot D)/\log\log(C \cdot D))$.

Corollary 7.7.1 and Corollary 5.2.2 together yield the following result.

Corollary 7.7.2. *For any network G of size N with routing number $R = \Omega(\log N)$, the duplication protocol can be used to route any permutation in time*

$$O\left(\frac{\log R \cdot \log\log N}{\log\log R} \cdot R\right),$$

w.h.p., using buffers of size $\Theta(\log R/\log\log R)$.

Using the fact that expanders of size N have routing number $O(\log N)$ immediately yields the following corollary.

Corollary 7.7.3. *For any bounded degree expander with N processors, N packets, one per processor, can be routed online according to an arbitrary permutation in*

$$O\left(\frac{\log N \cdot (\log\log N)^2}{\log\log\log N}\right)$$

time, w.h.p., using buffers of size $\Theta(\log\log N/\log\log\log N)$.

7.7.3 Limitations

Note that non-simple path systems cause problems for the duplication protocol, since in this case a duplication of the copies for each active packet might increase the congestion and therefore the collision probability by more than a factor of two.

7.8 The Protocol by Rabani and Tardos

In the following we present another protocol for routing packets along an arbitrary simple path collection of size n with congestion C and dilation D, using buffers of size $poly(\log n)$. It has been presented in [RT96].

7.8.1 Description of the Protocol

The main idea of the protocol follows the lines in the papers of Leighton *et al.* [LMR88, LMR94]. Instead of using the Lovász Local Lemma for refining a routing schedule, Rabani and Tardos developed techniques to guarantee the delivery of packets w.h.p. by using a simple version of Chernoff bounds for events with limited correlation. During the execution of their protocol it might happen that packets get delayed more than allowed by some (randomly chosen) schedule. In this case some special time is reserved for packets that are "far" behind where they are "supposed to be" in the schedule. This extra reserved time allows the delayed packets to catch up and continue on their paths, where they were "supposed to be".

Theorem 7.8.1. *Given any simple path collection $\mathcal{P}$ of size n with congestion C and dilation D, the protocol by Rabani and Tardos requires*

$$O(C) + (\log^* n)^{O(\log^* n)} D + poly(\log n)$$

time, w.h.p., to finish routing in $\mathcal{P}$, using buffers of size $poly(\log n)$.

Since the proof of this theorem is quite complicated, we refer to [RT96] and do not present it here. Instead, we present a proof of the following theorem, which contains many of the basic building blocks for the proof of Theorem 7.8.1 (see also [RT96]).

Theorem 7.8.2. *Given any simple path collection $\mathcal{P}$ of size n with congestion C and dilation D, there is a protocol that requires*

$$O(C + D \log \log n + poly(\log n))$$

time, w.h.p., to finish routing in $\mathcal{P}$, using buffers of size $poly(\log n)$.

Proof. Let us assume in the following that $C \leq D \log n$. Otherwise, the random delay protocol in Section 7.1 can already be used to obtain a runtime of $O(C)$. Furthermore, we assume that $D \leq n$. Otherwise, we choose the strategy of the random delay protocol to cut paths into pieces of length n and use the algorithm below for each round instead of Route(ℓ). As we will see, this ensures a runtime of $O(C + D \log \log n)$.

Before we start with a proof for the remaining cases, let us introduce some terminology. A *t-frame* is defined as a sequence of t consecutive time steps that starts at an integer multiple of t. Given a t-frame F, the *F-segment* of a scheduled path is defined as the part of it that is scheduled within F. Given a schedule S, the *contention* at edge e in S is defined as the maximum number of packets that want to pass e at the same time step.

Our strategy for constructing a protocol with the time bound above is to make a succession of refinements to an initial schedule S_0, in which each packet is forwarded at every time step until it reaches its final destination. The first step is to assign an initial delay to each packet, chosen independently and uniformly at random from $[2C/\nu]$, where $\nu = \log \log n$. Then the following lemma can easily be shown with the help of Chernoff bounds.

Lemma 7.8.1. *For any constant $c > 0$ there is a constant α such that with probability at least $1 - n^{-c}$ the congestion in any T-frame with $T = \alpha\nu \log n$ is at most νT.*

Let S_1 be a schedule that fulfills Lemma 7.8.1 with a suitably chosen α, depending on the probability bounds needed later. Our aim is to show that each T^6-frame in S_1 can be scheduled independently from the other frames in time $O(\nu)T^6$, w.h.p. This produces a schedule of total length

$$\left(\frac{D + 2C/\nu}{T^6} + 1\right) \cdot O(\nu T^6) = O\left(C + D \log \log n + T^6 \log \log n\right) ,$$

w.h.p., as desired. Let us therefore consider some fixed T^6-frame F in the following. Consider a T-frame f of F. We attempt to schedule all path segments in f by inserting an initial random delay from $[T]$ in front of every segment. Then the expected number of packets crossing an edge at any time step is bounded by ν. Hence the following lemma can easily be shown by using Chernoff bounds.

Lemma 7.8.2. *For any edge e and T-frame f in F it holds that for any constant $c > 0$ there is a constant α for T such that with probability at least $1 - T^{-c}$ the contention at e in f is at most 2ν.*

Consider, for some T-frame f, the schedule for f after inserting the initial random delays. We attempt to schedule all f-segments in this new schedule in an interval of length $2T \cdot 2\nu$ by allowing 2ν steps for the simulation of each step. Let an edge be called *blocked* in f if it does not fulfill Lemma 7.8.2 for f. An f-segment is called *successful* if it does not cross an edge that is blocked in f. Otherwise the f-segment *fails*. From Lemma 7.8.2 we get that the probability of an F-segment to fail in some T-frame in F is bounded by an inverse polynomial in T. However, we would like each F-segment to fail only with probability bounded by an inverse exponential in T. This is where we have to use a catch-up track to help packets catch up with the schedule after they fail. The key feature of the catch-up track is that the congestion there is much lower with sufficiently high probability, as the following analysis shows.

First we bound the number of failed F-segments that cross any particular edge e. Note that this includes not only segments that fail at e, but also all segments that were supposed to cross e, but fail due to some edge e'. Hence the failure of different F-segments through e may be highly correlated. In order to cope with the correlations, we introduce the following lemma, which extends the Chernoff bounds.

Lemma 7.8.3. *Assume that the maximum degree in the dependency graph of N binary random variables $X_1, \ldots, X_N$ is at most k, and suppose that for all i, $E[X_i] \le p$. Let $X = \sum_{i=1}^{N} X_i$ and $\delta \ge 1$. Then,*

$$\Pr[X > 4e\delta pN] < 4ek2^{-\delta pN/k} .$$

Proof. By Theorem 3.3.2, the dependency graph can be partitioned into at most $k + 1$ independent sets, and therefore into $m \le 4ek$ independent sets, each of size at most $N/2ek$. Let the set sizes be $N_1, N_2, \ldots, N_m$. Let S_i denote the number of variables in set i that are 1. Set $\beta_i = \delta N/kN_i \geq 2e$. Using the Chernoff bounds (see Theorem 3.2.2), we get

$$\Pr[S_i > \beta_i p N_i] < \left(\frac{e}{\beta_i}\right)^{\beta_i p N_i} .$$

Summing over the sets and using $\beta_i N_i = \delta N/k$ gives the lemma. $\square$

With this lemma, we are able to prove the following lemma.

Lemma 7.8.4. *Assume that the congestion of each T-frame in F is bounded by νT. Then for any constant $c > 0$ there is a constant α for T such that the probability that more than $4e\nu T^5$ F-segments crossing e fail is bounded by n^{-c}.*

Proof. Clearly, there are at most νT^6 F-segments that want to cross e. Let $\mathrm{FAIL}(i, j)$ denote the event that F-segment i failed in T-frame j. The total number of such events is νT^{11}. Our aim is to bound the number of events $\mathrm{FAIL}(i, j)$ that are true. An upper bound for these events is clearly an upper bound on the number of failed F-segments that cross e, because if some F-segment i fails then at least one of its events $\mathrm{FAIL}(i, j)$ must be true.

First, let us bound the dependencies among the events. Since for each T-frame in F new delays are chosen independently at random, $\mathrm{FAIL}(i, j)$ is totally independent of any $\mathrm{FAIL}(i', j')$ if $j \neq j'$. Now consider some T-frame j. Two events $\mathrm{FAIL}(i, j)$ and $\mathrm{FAIL}(i', j)$ are correlated iff there is an edge in the j-segment of i and an edge in the j-segment of i', whose blocking events are correlated. There are at most νT j-segments through an edge, and each j-segment goes through at most T edges. Hence, the blocking of an edge is correlated with at most νT^2 other edges. So a j-segment goes through T edges, each of these edges is correlated with νT^2 other edges, and each of these other edges carries at most νT other j-segments. This gives a total of $\nu^2 T^4$ j-segments whose failure correlates with the failure of a particular j-segment. Hence the dependency graph over the $\mathrm{FAIL}(i, j)$ events has maximum degree $\nu^2 T^4$.

Now we can apply Lemma 7.8.3. In our case, $N = \nu T^{11}$, the failure probability p can be made $1/T^{-c}$ for any constant c, and $k = \nu^2 T^4$. Let c be chosen such that $pN = T^5$. Let the random variable X denote the number of $\mathrm{FAIL}(i, j)$ events that are true. Then, for $\delta = \nu$,

$$\Pr[X > 4e\delta T^5] < 4e\nu^2 T^4 \cdot 2^{-\frac{\nu T^5}{\nu^2 T^4}} = 4e\nu^2 T^4 \cdot 2^{-T/\nu} ,$$

which can be made as small as $1/n^{-c}$ for any constant $c > 0$, depending on the choice of α in T. $\square$

Now we are ready to continue the schedule for F. Successful F-segments wait till the end of the frame. The rest of the scheduling process within F

handles failed F-segments. According to Lemma 7.8.4, at most $4e\nu T^5$ failed F-segments cross an edge, w.h.p. We schedule these segments as follows.

Partition F into T consecutive T^5-frames. This partitions the failed F-segments into T parts. We schedule these parts frame by frame by inserting random delays from $[T^5]$. In this schedule, blocked edges and failed segments are defined similar to above, but this time for the longer T^5-frames. If a path segment in a T^5-frame succeeds, the packet moves on to the next T^5-frame, trying to schedule the next path segment. If a path segment fails, it is moved to the next T^5-frame, where it tries again. Note that, according to Lemma 7.8.4, in any attempt to schedule T^5-frame path segments at most $4e\nu T^5$ segments cross any particular edge. Hence we can easily show the following lemma.

Lemma 7.8.5. *Let p be a path segment attempted to be scheduled in an iteration of the above scheduling process. For any constant $c > 0$, the probability that it fails can be made to be at most T^{-c} independent of all other events in all previous iterations.*

Once a packet succeeds T times (i.e., T consecutive T^5-frame path segments of this packet have been scheduled), it stops and waits till the end of the schedule for F. It remains to bound the number of trials a packet has to perform w.h.p. This will be done by the following lemma, which can easily be shown.

Lemma 7.8.6. *For any constant $c > 0$ there is a constant α for T such that the probability that a packet will succeed less than T times after $2T$ trials is bounded by n^{-c}.*

Hence the time needed to schedule the failed path segments is bounded by $O(T \cdot \nu T^5) = O(\nu T^6)$, which completes the proof. □

7.8.2 Applications

Together with Lemma 5.1.1 and Corollary 5.2.1, Theorem 7.8.1 implies the following result.

Theorem 7.8.3. *For any network G of size N with routing number $R \geq \log^k N$, $k \geq 0$ sufficiently large, any h-relation can be routed online in time $(O(h) + (\log^* N)^{O(\log^* N)})R$, w.h.p.*

7.8.3 Limitations

As for the offline protocols described in Chapter 6, the protocol by Rabani and Tardos requires the path collection to consist solely of simple paths.

7.9 The Protocol by Ostrovsky and Rabani

Let us conclude the list of oblivious routing protocols with presenting the asymptotically best protocol known so far for routing packets along an arbitrary simple path collection of size n with congestion C and dilation D. It has been presented in [OR97].

7.9.1 Description of the Protocol

The main idea of the protocol is based on the protocol by Rabani and Tardos, that is, during the execution of their protocol it might happen that packets get delayed more than allowed by some (randomly chosen) schedule. In this case again some special time is reserved for packets that are "far" behind where they are "supposed to be" in the schedule. This extra reserved time allows the delayed packets to catch up and continue on their paths, where they were "supposed to be". However, Ostrovsky and Rabani get rid of the factor $(\log^* n)^{O(\log^* n)}$ due to the recursive refinement procedure in [RT96] by collecting packets into much longer catch-up tracks, which have very small congestion relative to their length. Thus they are able to pack all the catch-up tracks together, and avoid recursion. However, the large catch-up tracks would introduce an additive term which is even larger than the additive term of $poly(\log n)$ in the protocol by Rabani and Tardos. To avoid it, they introduce an initial refinement process that is invoked a constant number of times, and avoids the large additive term by decomposing the collection of failed packets into small connected components (with respect to correlations) w.h.p. The idea of this process is based on the first step in Beck's algorithmic version of the Lovász Local Lemma [Be91]. The small size of each component makes it behave, in some sense, like a single edge. Thus Ostrovsky and Rabani are able to schedule each component separately by simply repeatedly trying to schedule them independently from each other until all packets succeed.

Theorem 7.9.1. *Given any simple path collection $\mathcal{P}$ of size n with congestion C and dilation D, the protocol by Ostrovsky and Rabani requires*

$$O\left(C + D + \log^{1+\epsilon} n\right)$$

time for any constant $\epsilon > 0$, w.h.p., to finish routing in $\mathcal{P}$, using buffers of size $poly(\log n)$.

Since the proof of this theorem is quite complicated, we refer to [RT96] and do not present it here.

7.9.2 Applications

Together with Lemma 5.1.1 and Corollary 5.2.1, Theorem 7.9.1 implies the following result.

Theorem 7.9.2. *For any network G of size N with routing number R, any h-relation can be routed online in time $O(h \cdot R + \log^{1+\epsilon} N)$, w.h.p., for any constant $\epsilon > 0$.*

7.9.3 Limitations

As for the protocol by Rabani and Tardos, the protocol by Ostrovsky and Rabani requires the path collection to consist solely of simple paths.

7.10 Summary of Main Results

In this section, we summarize the main results we presented about oblivious routing in fixed path collections and networks. We presented several protocols for different types of path collections. After we described and analyzed a very simple, non-optimal protocol for arbitrary path collections, we showed the following result (see Theorem 7.2.1).

For any leveled path collection $\mathcal{P}$ of size n with congestion C and depth D, the random rank protocol needs at most $O(C + D + \log n)$ time steps to finish routing in $\mathcal{P}$, w.h.p., using edge buffers of size C.

It seems that the random rank protocol can only be efficiently used for arbitrary leveled path collections with sufficiently large buffers, since there exist simple path collections (see Observation 7.2.1) where the expected routing time of the random rank protocol is bounded by $\Omega((\log n/\log\log n)^{1/2}(C + D))$. Furthermore, there exist leveled path collections for which the runtime of the random rank protocol for buffers of size one is bounded by $\Omega(C \cdot D)$ (see Observation 7.2.2). The latter drawback is removed by the protocol we presented next (see Theorem 7.3.1).

For any leveled path collection $\mathcal{P}$ of size n and bounded degree with congestion C and depth D, Ranade's protocol needs at most $O(C + D + \log n)$ time steps to finish routing in $\mathcal{P}$, w.h.p., using edge buffers of size one.

Ranade's protocol can be efficiently applied to routing in wrap-around butterflies, shuffle-exchange networks, and d-dimensional meshes and tori. For Ranade's protocol to be optimal it is important that the depth of a path collection is at most a constant away from its dilation. Otherwise path collections can be constructed that have an expected runtime of $\Omega(\log^2 n/\log\log n)$ even for constant dilation and congestion (see Observation 7.3.1). The next protocol we presented is especially useful for arbitrary shortcut-free path collections (see Theorem 7.4.1).

For any d-shortcut-free path collection $\mathcal{P}$ of size n with congestion C and dilation D, the growing rank protocol needs at most $O(C+\max\{1, \frac{\log(nD)}{d}\}D)$ time steps to finish routing in $\mathcal{P}$, w.h.p., using edge buffers of size C.

The growing rank protocol can be efficiently applied to routing in arbitrary node-symmetric networks and networks with girth at least logarithmic in their size. (Note that expander graphs of size N exist with girth $\Omega(\log N)$, see e.g. [LPS88].) For constant d we could show that there exist simple path collections for which the expected runtime of the growing rank protocol is bounded by $\Omega(C + D \cdot \log n / \log\log n)$. This demonstrates that the upper bound above is nearly tight. The growing rank protocol can be generalized to the extended growing rank protocol. For this protocol, we could show the following result.

For any simple path collection $\mathcal{P}$ of size n that can be decomposed into S stages of shortcut-free path collections with congestion C and dilation D, the extended growing rank protocol needs at most $O(S(C + D) + \log n)$ time steps to finish routing in $\mathcal{P}$, w.h.p., using edge buffers of size $S \cdot C$.

The extended growing rank protocol can be efficiently applied to network emulations. In particular, we could show that for any node-symmetric network H of size N with diameter $D = \Omega(\log N)$ and any bounded degree network G of size N with routing number R, any permutation routing problem in G can be simulated by H in time $O(R \cdot D)$, w.h.p.

Both the growing rank protocol and extended growing rank protocol can deadlock if the buffer size is too small. Therefore, we presented two protocols that can even cope with small buffers. The first result can be found in Corollary 7.6.1, and the second result is due to Corollary 7.7.1.

For any simple path collection $\mathcal{P}$ of size n with link buffers of size B, congestion C and dilation D, the trial-and-failure protocol needs at most $O(C \cdot D^{1/B} + D \log n)$ time steps to finish routing in $\mathcal{P}$, w.h.p.

For any simple path collection $\mathcal{P}$ of size n with congestion C and dilation D, the duplication protocol needs at most

$$O\left(\left(C + D\log\log n + \frac{\log n \cdot \log\log n}{\log\log(C \cdot D)}\right) \cdot \frac{\log(C \cdot D)}{\log\log(C \cdot D)}\right)$$

time steps to finish routing in $\mathcal{P}$, w.h.p., using buffers of size $\Theta(\log(C \cdot D)/\log\log(C \cdot D))$.

For C or D at least $\log^{1+\epsilon} n$ for some constant $\epsilon > 0$, the best protocol known so far for arbitrary simple path collections with sufficiently large buffer size is the protocol by Ostrovsky and Rabani (see Theorem 7.9.1).

Given any simple path collection $\mathcal{P}$ of size n with congestion C and dilation D, the protocol by Ostrovsky and Rabani requires

$$O\left(C + D + \log^{1+\epsilon} n\right)$$

time for any constant $\epsilon > 0$, w.h.p., to finish routing in $\mathcal{P}$, using buffers of size poly$(\log n)$.

This result implies that for any network of size N with routing number $R \geq \log^{1+\epsilon} N$ for some constant $\epsilon > 0$ there exists an (average case) optimal online protocol for routing arbitrary permutations.

8. Adaptive Routing Protocols

We start this chapter with a motivation why adaptive routing protocols are important. Afterwards, we present a network in which any s-relation can be routed deterministically in optimal time. This network is called extended s-ary multibutterfly. Furthermore, we describe techniques for developing universal adaptive protocols and give an overview about what has been found out in this area so far. One of these techniques is called "routing via simulation". Together with the protocol for routing arbitrary s-relations in s-ary multibutterflies it is used to construct efficient deterministic routing protocols for arbitrary networks.

Adaptive protocols are very appealing compared to oblivious protocols, since they allow a parallel system to react more flexibly in case of faulty or overloaded communication links or processors. Another motivation for adaptive routing is that bounded buffers are difficult to handle for oblivious routing strategies. Usually, the only way to avoid deadlocks using oblivious routing strategies is simply to delete the packets in case of full buffers and restart them again from the source (see, e.g., the trial-and-failure protocol). In the deterministic case, adaptive routing protocols are also usually far superior against oblivious routing protocols, since the worst case congestion for routing an arbitrary permutation using a fixed path system can be fairly large (see Theorem 5.2.1). But even if the congestion is small, deterministic oblivious routing strategies might still perform very poorly as will be shown in the following.

Let a routing protocol be called *non-predictive* if contention is resolved by a deterministic algorithm that is based only on the history of the contending packet's travels through the network and on information carried with the packets that is independent of their destination. For example, the random rank protocol is not deterministic and hence not non-predictive. However, for any fixed setting of the initial ranks it is non-predictive. The same holds for the growing rank protocol and the extended growing rank protocol. The following example shows that all non-predictive protocols perform poorly even on leveled networks. The proof has been given by Maggs and Sitaraman [MS92].

Theorem 8.0.1. *For any deterministic non-predictive protocol, there exists a permutation π that requires $\Omega(N/q \log N)$ steps to route on a butterfly of size N with buffers of size q although its congestion is only $O(q \log N)$.*

Note that this result also holds for a deterministic version of Ranade's protocol. Therefore in case of deterministic routing, efficient adaptive routing protocols are highly needed. We start this section with describing networks for which optimal deterministic routing protocols are already known.

8.1 Deterministic Routing in Multibutterflies

For s-ary multibutterflies of size N it is known that any permutation can be routed deterministically in time $O(\log_s N)$ (see Section 4.2.2). In this section we extend the definition of the s-ary multibutterfly such that it can be used to route any s-relation in optimal time.

8.1.1 The r-replicated s-ary Multibutterfly

The basic building block of the (1-replicated) s-ary multibutterfly is an s-ary m-router.

The s-ary m-router (or (s, m)-router) is a bipartite graph with m input nodes and m output nodes. It is a combination of the following two graphs.

The first graph is called s-ary m-distributor. It is a directed graph with all input nodes as node set. Each input node is the starting point of s edges numbered from 1 to s. We require the edges to be chosen such that the set of all endpoints of the ith edges forms a permutation, $i \in \{1, \ldots, s\}$, and specific expansion properties described later are fulfilled. This graph will be used to balance the distribution of the packets in the input nodes.

We call the second graph s-ary m-splitter (see Section 4.2.2). In this graph the output nodes are separated into $\sqrt{s}$ output sets, each with $m/\sqrt{s}$ nodes. Every input node has $\sqrt{s}/2$ edges to each of the $\sqrt{s}$ output sets. The edges connecting the input set to each of the output sets define an expander graph with properties we will describe later. This graph will be used to forward the packets to their destinations.

The s-ary d-dimensional multibutterfly (s, d)-MBF has d levels. The nodes at level $0 \leq i \leq d-1$ are partitioned into $\sqrt{s}^i$ sets of $m_i = \sqrt{s}^{d-i}$ consecutive nodes. Each of these sets in level i is an input set of an s-ary m_i-router. The output sets of that router are $\sqrt{s}$ sets of size m_{i+1} in level $i + 1$. Thus each node in the (s, d)-MBF is the endpoint of at most $2(s + \frac{s}{2}) = 3s$ edges.

In Section 8.2.2 we also need the notion of an (s, d, k)-MBF. This multibutterfly is defined as follows. Let the (s, m, k)-router be a graph with $k \cdot m/\sqrt{s}$ input nodes and k sets of $m/\sqrt{s}$ output nodes. The input nodes are connected by an s-ary $k \cdot m/\sqrt{s}$-distributor. The s-ary splitter of such a router is modified in a way that every input node has $\frac{s}{2k}$ edges to each of the k output sets.

We call such a splitter (s, m, k)-splitter. For $k \in \{1, \ldots, \sqrt{s}\}$, the (s, d, k)-MBF can be derived from an (s, d)-MBF by replacing the (s, m)-router at level 0 of the (s, d)-MBF by an (s, m, k)-router as shown in Figure 8.1.

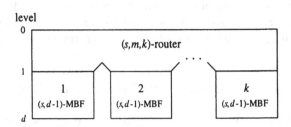

Fig. 8.1. The structure of an (s, d, k)-MBF.

The r-replicated s-ary multibutterfly (r, s, d, k)-MBF is defined by taking r copies of the (s, d, k)-MBF and identifying the corresponding nodes. That is, each edge now is able to forward r packets in one time step.

In the rest of this section we present a proof for the following theorem (see also [MS96a]).

Theorem 8.1.1. *Given an extended r-replicated s-ary multibutterfly of size N with $r \geq 1$ and $s \geq 2$, $r \cdot s \cdot N$ packets, $r \cdot s$ per processor, can be routed deterministically according to some arbitrary $r \cdot s$-relation in time $O(\log_s N)$.*

Proof. Let n denote the number of nodes in one level of a given r-replicated s-ary MBF. Consider first the simpler problem of routing $rs \cdot n$ packets, rs per node at the top level, to the bottom level of the MBF according to some arbitrary rs-relation. In order to do this time-efficiently we use the following protocol.

8.1.2 Description of the Simple Protocol

We partition the $rs \cdot n$ messages into 522 batches such that no more than $rs \cdot m/522$ messages from each batch traverse any m-router. Batch j has all packets with destinations in the set $\{z \mid z = j - 1 \pmod{522}\}$. For the purpose of analysis we assume that all batch j packets have been routed before transmitting batch $j + 1$ packets, and we now concentrate on the routing of one batch.

Nodes at even levels of the MBF transmit in odd phases, nodes at odd levels transmit in even phases (that is, level 0 is active in phase $1, 3, \ldots$). A phase has three subphases. Consider an (s, m, k)-router in which the input nodes want to transmit. We assume that no node has more than $r \cdot s$ packets.

The task of the *balancing phase* is to distribute the packets in such a way among the input nodes that there are only very few nodes with more than $\frac{r \cdot s}{2}$ packets. As we will see, this can be obtained if each input node sends out its $p \leq r \cdot s$ packets along the edges of the m-distributor with numbers 1 to $\lceil \frac{p}{r} \rceil$ in an arbitrary order. Note that, according to the definition of the m-distributor, each input node receives at most $r \cdot s$ packets.

The task of the *placement phase* is to distribute the packets in such a way among the input nodes that there are only very few nodes having more than $\frac{r \cdot s}{4k}$ packets that have to be sent to the same output set. As we will see, this can be obtained if the packets are distributed among the s edges of the m-distributor in a specific way, before sending them out.

The *delivery phase* consists of three steps. Its task is to send as many packets as possible to output sets prescribed by their destinations. For each input node, only the first $\frac{r \cdot s}{4k}$ packets to each of the k output sets are declared active. In the first step each input node sends for each output set to which it has messages to transmit a request message along all $\frac{s}{2k}$ edges of the m-splitter to that output set. An output node that currently stores at most $\frac{r s}{2}$ packets replies in the second step with a ready message to all input nodes that sent a request to it. In the third step each input node sends up to r active packets to each node that sent it a ready message.

8.1.3 Analysis of the Simple Protocol

We first analyze the routing of one batch. Consider an (s, m, k)-router in a phase in which the inputs of that router are active. Let x (resp. x') be the total number of packets that are stored in the input nodes of that splitter at the beginning (resp. end) of that phase. Let y be the number of packets stored in output nodes of the splitter at the beginning of this phase.

Lemma 8.1.1.

$$x' \leq \frac{400 \log s}{\sqrt{s}} (x + y) \ .$$

Proof. Both the balancing phase and the placement phase require the packets to be distributed among the edges of the m-distributor in a suitable way. For each input node v, this will be done with the help of an assignment graph that is defined as follows.

Assume that every packet has one out of c possible colors. The *assignment graph* $A = (P_A, D_A, E_A)$ is defined as a bipartite undirected graph with node sets P_A and D_A, and a set of edges E_A. $P_A = \{v_{i,j} \mid i \in \{1, \ldots, c\}, j \in \{1, \ldots, s\}\}$ consists of $c \cdot s$ nodes, s nodes for each color, and D_A consists of s nodes representing all s edges leaving v in the m-distributor. Each node in P_A has d edges to nodes in D_A. Consider v to have p_i packets with color i, partitioned into $\lceil \frac{p_i}{r} \rceil$ *blocks* of size at most r. We allow the jth of these blocks to be assigned only to nodes in the set $\Gamma(\{v_{i,j}\})$. (For any node set

U, $\Gamma(U)$ is defined as the set of all nodes v that are adjacent to a node $u \in U$.) In case that under this restriction every block can be assigned to a different node in D_A, each block is sent along that edge of the m-distributor that is represented by the node in D_A that has been assigned to the block. If every input node uses the same assignment graph then, as will be shown in Proposition 8.1.4, the edges of the m-distributor can be distributed in such a way among the input nodes that, after sending the blocks of packets along the edges assigned to them, the distribution of the packets among the inputs is close to be balanced w.r.t any color.

For the balancing phase an assignment graph can easily be constructed. The task of this phase is to distribute the packets among the input nodes in such a way that there is only a very small portion of packets left that is stored in nodes with more than $\frac{r \cdot s}{2}$ packets. Hence we need only one color. Consider input v to have p packets, partitioned into $\lceil \frac{p}{r} \rceil$ blocks of size at most r. Then we simply assign the jth edge of the m-distributor to the jth of these blocks. The underlying assignment graph has degree $d = 1$. Consider marking the first (at most) $\frac{r \cdot s}{2}$ packets in each input node after the balancing phase. Then we are interested in the total number of unmarked packets. An upper bound for this number will be given in the following proposition.

Proposition 8.1.1. *There exists an s-ary m-distributor such that after the balancing phase at most $\frac{48x \log s}{s}$ unmarked packets are stored in the input nodes.*

Proof. The result follows from Proposition 8.1.4 with $d = 1$, $c = 1$, $z = \frac{s}{2}$ and $\epsilon \leq \frac{1}{522} \leq \frac{c \cdot z}{12 ed \cdot s}$. □

For the placement phase it is more difficult to find a suitable assignment graph. The task of the placement phase is to distribute the packets in such a way among the input nodes that there are only very few nodes having more than $\frac{r \cdot s}{4k}$ packets that have to be sent to the same output set. Hence we need k colors representing the k output sets of the (s, m, k)-router. If we only concentrate on assigning edges of the m-distributor to marked packets, we can prove the following result.

Proposition 8.1.2. *There is an assignment graph A of degree 4 such that, for any choice of colors for the at most $\frac{r \cdot s}{2}$ marked packets stored in an input node, A can be used to assign at most one block of marked packets to each edge such that the jth block of color $i \in \{1, \ldots, k\}$ is assigned to an edge in $\Gamma(\{v_{i,j}\})$.*

Proof. Consider a fixed input node v. The proof consists of a probabilistic argument. In particular, we show that for randomly chosen endpoints in D_A for the edges in E_A and suitably chosen d the probability that the resulting graph A does not fulfill the proposition is smaller than one. Therefore, there exists an assignment graph A such that the marked packets can be distributed among the edges in such a way that each edge gets at most one block of packets.

Let $p \leq k$ denote the number of colors used by the marked packets in v, let i_j be the number of blocks of marked packets in v assigned to the jth of these p colors. Since each input node stores at most $\frac{r \cdot s}{2}$ marked packets, the number q of blocks is at most $\sum_{j=1}^{p} i_j \leq \frac{s}{2} + p$ (for each of the p colors there may exist a block with less than r packets). There are at most $\binom{s}{q-1}$ possibilities for choosing a subset of size $q - 1$ out of s possible nodes in D_A. The probability that all alternatives for the blocks point to nodes within such a subset is bounded by $(\frac{q-1}{s})^{d \cdot q}$. Then the probability that for any choice of colors for the packets there is a subset S of nodes in P_A representing blocks of packets with $\Gamma(S) < |S|$ is bounded above by

$$\sum_{p=1}^{k} \binom{k}{p} \sum_{\substack{1 \leq i_1, \ldots, i_p \leq s, \\ \sum_j i_j \leq s/2+p}} \binom{s}{\sum_j i_j - 1} \left(\frac{\sum_j i_j - 1}{s}\right)^{d \sum_j i_j}$$

$$\leq \sum_{p=1}^{k} \binom{k}{p} \sum_{q=p}^{s/2+p} \binom{q-1}{p-1} \binom{s}{q-1} \left(\frac{q-1}{s}\right)^{d \cdot q} \qquad (8.1)$$

In order to simplify the formula, we need the following claim.

Claim 8.1.1. *For a sufficiently large s and $d \geq 4$,*

$$\binom{s}{p} \left(\frac{p}{s}\right)^{d \cdot p} \geq \binom{s}{q-1} \left(\frac{q-1}{s}\right)^{d \cdot q}$$

for all $p \in \{1, \ldots, k\}$ and $q \in \{p, \ldots, s/2 + p\}$.

Proof. It holds

$$\binom{s}{q} \left(\frac{q}{s}\right)^{d \cdot q} \geq \binom{s}{q+1} \left(\frac{q+1}{s}\right)^{d(q+1)}$$

$$\Leftrightarrow \quad \frac{\binom{s}{q}}{\binom{s}{q+1}} \geq \left(1 + \frac{1}{q}\right)^{d \cdot q} \left(\frac{q+1}{s}\right)^{d}$$

$$\Leftarrow \quad \frac{q+1}{s-q} \geq \left(\frac{e(q+1)}{s}\right)^{d},$$

which is true for all $d \geq 4$ and $q \in \{p, \ldots, \frac{s}{2e} - 1\}$, if s is sufficiently large. Hence $\binom{s}{p}(\frac{p}{s})^{d \cdot p} \geq \binom{s}{q-1}(\frac{q-1}{s})^{d \cdot q}$ for all $q \in \{p+1, \ldots, \frac{s}{2e} - 1\}$. In case that $q \in \{\frac{s}{2e}, \ldots, \frac{s}{2} + p\}$ and s sufficiently large (recall that $p \leq \sqrt{s}$), we get

$$\binom{s}{q-1} \left(\frac{q-1}{s}\right)^{d \cdot q} \leq e^q \left(\frac{q-1}{s}\right)^{(d-1)q} \leq \left(\frac{1}{2}\right)^q$$

$$\leq \left(\frac{p}{s}\right)^{(d-1)p} \leq \binom{s}{p} \left(\frac{p}{s}\right)^{d \cdot p}.$$

In case that $q = p$ and $p > 1$ (the case $p = 1$ is obviously correct) we get

$$\binom{s}{p}\left(\frac{p}{s}\right)^{d \cdot p} \geq \binom{s}{p-1}\left(\frac{p-1}{s}\right)^{d \cdot p}$$

$$\Leftrightarrow \quad \left(\frac{p}{p-1}\right)^{d \cdot p} \geq \frac{p}{s-p+1} \; ,$$

which is also true. Hence the claim follows.   □

Let $d \geq 4$ and s be sufficiently large (note that $k \leq \sqrt{s}$). Then it follows from Claim 8.1.1 that

$$(8.1) \quad \leq \sum_{p=1}^{k} \binom{k}{p}\binom{s}{p}\left(\frac{p}{s}\right)^{d \cdot p} \sum_{q=p}^{s/2+p} \binom{q-1}{p-1}$$

$$\leq \sum_{p=1}^{k} \binom{k}{p}\left(\frac{es}{p}\right)^{p}\left(\frac{p}{s}\right)^{d \cdot p}\binom{s/2+p}{p}$$

$$\leq \sum_{p=1}^{k} \binom{k}{p}\left(\frac{e(s/2+p)}{p}\right)^{p} e^{p}\left(\frac{p}{s}\right)^{(d-1)p}$$

$$\leq \sum_{p=1}^{k} \left(\frac{ek}{p}\right)^{p} e^{2p}\left(\frac{p}{s}\right)^{(d-2)p}$$

$$\leq \sum_{p=1}^{k} \left(\frac{p}{s}\right)^{(d-3)p} < 1 \; .$$

Using Hall's matching theorem (see Theorem 3.3.1) it follows that the probability is smaller than one that, for randomly chosen endpoints in D_A for the edges in E_A, there exists a coloring of the marked packets such that an assignment of marked packets to nodes in D_A as described above is not possible. Thus the proposition holds.   □

Let each edge of the distributor represent r channels. Then the unmarked packets will be assigned to those of the $r \cdot s$ channels that are not used by the marked packets. Although we do not consider the unmarked packets in the following, it is important to let them participate in the placement phase to ensure that after the placement phase every input node has at most $r \cdot s$ packets. Let us call the first $\frac{r \cdot s}{4k}$ marked packets for each output set that are stored in each input node after the placement phase *active*. Then we can prove the following proposition.

Proposition 8.1.3. *There exists an s-ary m-distributor such that at most $\frac{384x \log s}{\sqrt{s}}$ marked packets are not active at the end of the placement phase.*

Proof. The result follows from Proposition 8.1.4 with $d = 4$, $c = k \leq \sqrt{s}$, $z = \frac{s}{4k}$, and $\epsilon \leq \frac{1}{522} \leq \frac{c \cdot z}{12ed \cdot s}$.   □

It remains to prove that, for a given assignment graph A, there exists an m-distributor such that the packets are close to be balanced among the input nodes w.r.t. any color.

Proposition 8.1.4. *Let A be any assignment graph for c colors with degree d that can be used to assign every edge of the m-distributor to at most one block of packets of any color in every input node. Consider marking for each color the first z packets in each input node after the packets have been sent along the edges of the m-distributor. If there are at most $\frac{\epsilon rsm}{c}$ packets of each color stored in the input nodes, $z \geq \sqrt{s}/4$, $\epsilon \leq \frac{c \cdot z}{12ed \cdot s}$, and s sufficiently large, then there exists an s-ary m-distributor such that the total number of unmarked packets is at most*

$$\frac{24d \cdot x \log s}{z} .$$

Proof. We use a probabilistic proof to show that there is a suitable distribution of the edges of the m-distributor among the input nodes such that Proposition 8.1.4 holds.

Consider a fixed color $\gamma \in \{1, \ldots, c\}$. Let x_γ denote the number of packets stored in the input nodes that have color γ. Let $p \geq z$, $q = \frac{p}{2d}$, and b_i denote the number of input nodes that have at least i blocks of packets with color γ, $i \in \{q, \ldots, \frac{s}{d}\}$. Note that if an input node has i blocks of packets with color γ then it has at least $f(i) = (i-1)r + 1$ packets with color γ. Thus the maximal number of input nodes with $i > q$ blocks given x_γ and $b_q, \ldots, b_{i-1}$ is at most

$$\min \left\{ \left\lceil \frac{x_\gamma - g(b_q, \ldots, b_{i-1})}{r} \right\rceil, b_{i-1} \right\}$$

with $g(b_q, \ldots, b_{i-1}) = f(q) \cdot b_q + \sum_{j=q+1}^{i-1} r \cdot b_j$. The following claim reveals why the assignment graph is so important for balancing the packets among the input nodes.

Claim 8.1.2. *If all input nodes use the same assignment graph A of degree d then the s edges reaching any input v can be numbered from 1 to s such that a block of packets with color γ that arrives at v via edge j implies at least $\lfloor \frac{j}{d} \rfloor$ blocks of packets of color γ in its origin before the packets have been sent out.*

Proof. Let all input nodes use the same assignment graph. Then it holds that, since for every $j \in \{1, \ldots, s\}$ the endpoints of the jth edges from all input nodes form a permutation, each input node is the endpoint of s edges representing the complete node set D_A. Consider the edges to be numbered from 1 to s such that if edge e has a lower number then edge e' then $\min\{j \mid e \in \Gamma(\{v_{\gamma,j}\})\} \leq \min\{j \mid e' \in \Gamma(\{v_{\gamma,j}\})\}$. Since the jth block of color γ

is only allowed to choose among d nodes in D_A it follows that if a block of color γ reaches a node v via edge j then its origin must have had at least $\lfloor \frac{i}{d} \rfloor$ blocks of packets with color γ.   □

With the help of this claim we can prove the following claim.

Claim 8.1.3. *The probability that after the placement phase at least u input nodes have at least p blocks of packets, each, with color γ is bounded by*

$$\binom{m}{u} \binom{m}{\frac{x_\gamma}{f(p/2d)}} \left(\frac{s}{d}\right)^{\frac{x_\gamma}{f(p/2d)}} \left(\frac{2ed \cdot x_\gamma}{p \cdot r \cdot m}\right)^{\frac{u \cdot p}{2}} .$$

Proof. For randomly chosen endpoints of the edges of the m-distributor, the probability that after the placement phase at least u input nodes have at least p blocks of packets, each, with color γ is bounded by

$$\binom{m}{u} \sum_{\substack{b_q,\ldots,b_{s/d} \geq 0, \\ \forall i : b_i \geq b_{i+1}, \ g(b_q,\ldots,b_{s/d}) \leq x_\gamma}} \left[\sum_{\frac{p}{2} \leq i_1 < \ldots < i_{p/2} \leq s} \prod_{j=1}^{p/2} \frac{b_{\lfloor i_j/d \rfloor}}{m} \right]^u . \tag{8.2}$$

This formula is derived as follows.

- There are $\binom{m}{u}$ ways to choose a set U of input nodes of size u.
- If p blocks of packets are sent to input node v, there must exist at least $\frac{p}{2}$ blocks that use an edge set $\{i_1, \ldots, i_{p/2}\} \subseteq \{\frac{p}{2}, \ldots, s\}$ with numbers chosen as defined in Claim 8.1.2. This entails a probability of $\frac{b_{\lfloor i_j/d \rfloor}}{m}$ that the i_jth edge is used by a block with color γ. Thus we get that, for each input node v in U, the probability that it gets at least p blocks of marked packets with color γ is at most

$$\sum_{\frac{p}{2} \leq i_1 < \ldots < i_{p/2} \leq s} \prod_{j=1}^{p/2} \frac{b_{\lfloor i_j/d \rfloor}}{m} . \tag{8.3}$$

Since there is only a limited number of edges that have their origin in nodes with at least q packets of color γ, the probabilities for the nodes in U are negatively correlated and therefore can be regarded as independent for an upper bound. This means that we get an overall probability of at most $(8.3)^u$.

Because of $f(q) \cdot b_q + \sum_{j=q+1}^{s/d} r \cdot b_j \leq x_\gamma$ (see the formula for g above) it holds that

$$\sum_{j=\frac{p}{2}}^{s} b_{\lfloor j/d \rfloor} \leq d \cdot \sum_{j=\frac{p}{2d}}^{s/d} b_j = d \left(b_q + \frac{1}{r} \sum_{j=q+1}^{s/d} r \cdot b_j \right)$$

$$\leq d \left(b_q + \frac{1}{r} (x_\gamma - f(q) b_q) \right) \leq \frac{d \cdot x_\gamma}{r} .$$

Thus we get

$$(8.3) \quad \leq \quad \frac{1}{(p/2)!} \sum_{i_1,\ldots,i_{p/2}\in\{\frac{p}{2},\ldots,s\}} \prod_{j=1}^{p/2} \frac{b_{\lfloor i_j/d \rfloor}}{m}$$

$$= \frac{1}{(p/2)!} \sum_{i_1,\ldots,i_{p/2-1}\in\{\frac{p}{2},\ldots,s\}} \prod_{j=1}^{p/2-1} \frac{b_{\lfloor i_j/d \rfloor}}{m} \sum_{i_{p/2}\in\{\frac{p}{2},\ldots,s\}} \frac{b_{\lfloor i_j/d \rfloor}}{m}$$

$$\leq \frac{1}{(p/2)!} \sum_{i_1,\ldots,i_{p/2-1}\in\{\frac{p}{2},\ldots,s\}} \prod_{j=1}^{p/2-1} \frac{b_{\lfloor i_j/d \rfloor}}{m} \cdot \frac{d \cdot x_\gamma}{r \cdot m}$$

$$\leq \ldots \leq \frac{1}{(p/2)!} \left(\frac{d \cdot x_\gamma}{r \cdot m} \right)^{p/2} \leq \left(\frac{2ed \cdot x_\gamma}{p \cdot r \cdot m} \right)^{p/2}.$$

Using this in (8.2) we get that the probability that at least u input nodes have at least p blocks of marked packets, each, with color γ is bounded by

$$\binom{m}{u} \sum_{\substack{b_q,\ldots,b_s \geq 0, \\ \forall i: b_i \geq b_s/d,\ g(b_q,\ldots,b_s/d) \leq x_\gamma}} \left(\frac{2ed \cdot x_\gamma}{p \cdot r \cdot m} \right)^{\frac{u \cdot p}{2}}$$

$$\leq \binom{m}{u} \binom{m}{\frac{x_\gamma}{f(q)}} \left(\frac{s}{d} \right)^{\frac{x_\gamma}{f(q)}} \left(\frac{2ed \cdot x_\gamma}{p \cdot r \cdot m} \right)^{\frac{u \cdot p}{2}}.$$

Since $q = \frac{p}{2d}$, the claim follows. $\square$

Let $u_{x_\gamma,p} \cdot p = \frac{4x_\gamma \log s}{f(p/2d)}$ and s be sufficiently large. Then the probability that at least $u_{x_\gamma,p}$ output nodes have at least $p \geq z$ blocks of marked packets, each, with color γ is bounded by

$$\sum_{\substack{f(p/2d) \leq x_\gamma \leq \frac{\epsilon r s m}{c}, \\ f(p/2d) | x_\gamma}} \binom{m}{u_{x_\gamma,p}} \binom{m}{\frac{x_\gamma}{f(p/2d)}} \left(\frac{s}{d} \right)^{\frac{x_\gamma}{f(p/2d)}} \left(\frac{2ed \cdot x_\gamma}{p \cdot r \cdot m} \right)^{\frac{p \cdot u_{x_\gamma,p}}{2}}$$

$$= \sum_{v=1}^{\frac{\epsilon r s m}{c f(p/2d)}} \binom{m}{\frac{4v \log s}{p}} \binom{m}{v} \left(\frac{s}{d} \right)^v \left(\frac{ev}{m} \right)^{2v \log s}$$

$$\leq \sum_{v=1}^{\frac{\epsilon r s m}{c f(p/2d)}} \left(\frac{ep \cdot m}{4v \log s} \right)^{\frac{4v \log s}{p}} \left(\frac{e s m}{d \cdot v} \right)^v \left(\frac{ev}{m} \right)^{2v \log s}$$

$$\stackrel{p \geq \sqrt{s}/4}{\leq} \sum_{v=1}^{\frac{\epsilon r s m}{c f(p/2d)}} \left(\frac{m}{ev} \right)^v \left(\frac{m}{ev} \right)^{v(\log s - 1)} \left(\frac{ev}{m} \right)^{2v \log s}$$

$$\stackrel{\epsilon \leq \frac{c \cdot f(p/2d)}{4ers}}{\leq} \sum_{v=1}^{\frac{\epsilon r s m}{c f(p/2d)}} \left(\frac{1}{4} \right)^{v \log s} \leq \frac{2}{s^2}$$

Summing over all $p = z$ to s this yields a probability smaller than 1 if $s > 2$. Note that for $\frac{z}{2d} \geq 3$ we have

$$\epsilon \leq \frac{c \cdot z}{12ed \cdot s} \leq \frac{c(z/(2d) - 1)}{4es} \leq \frac{c \cdot f(p/2d)}{4ers} \ .$$

Since there exists an m-distributor with $u_{x_\gamma, p} \cdot p \leq \frac{4x_\gamma \log s}{f(p/2d)}$ for any distribution of the packets such that the assignment graph can be applied, at most $\frac{4d \cdot x_\gamma \log s}{p \cdot f(p/2d)}$ input nodes have at least $p \geq z$ blocks of marked packets after sending them along the edges of the distributor.

Let c_i be the number of input nodes that have exactly i blocks of marked packets after sending them along the edges of the distributor, $i \in \{z, \ldots, s\}$. Then the total number of unmarked blocks of packets is at most $\sum_{i=z}^{s} i \cdot c_i$. Since we require the c_i to obey

$$\sum_{i=p}^{s} c_i \leq \frac{4d \cdot x_\gamma \log s}{p \cdot f(p/2d)}$$

for all $p \geq z$, $\sum_{i=z}^{s} i \cdot c_i$ gets maximal if we set

$$c_s = \frac{4d \cdot x_\gamma \log s}{s \cdot f(s/2d)} \ ,$$

$$c_{s-1} = \frac{4d \cdot x_\gamma \log s}{(s - 1) \cdot f((s - 1)/2d)} - \frac{4d \cdot x_\gamma \log s}{s \cdot f(s/2d)} \ ,$$

$$c_{s-2} = \frac{4d \cdot x_\gamma \log s}{(s - 2) \cdot f((s - 2)/2d)} - \frac{4d \cdot x_\gamma \log s}{(s - 1) \cdot f((s - 1)/2d)} \ ,$$

$$\ldots$$

From this we conclude that for a sufficiently large s and $\frac{z}{2d} \geq 3$ the edges of the m-distributor can be chosen in such a way that at most

$$r \left[\sum_{p=z}^{s-1} p \left(\frac{4x_\gamma \log s}{p \cdot f(p/2d)} - \frac{4x_\gamma \log s}{(p + 1) \cdot f((p + 1)/2d)} \right) + s \cdot \frac{4x_\gamma \log s}{sf(s/2d)} \right]$$

$$\leq r \left[\sum_{p=z}^{s-1} p \left(\frac{12d \cdot x_\gamma \log s}{r \cdot p^2} - \frac{12d \cdot x_\gamma \log s}{r(p + 1)^2} \right) + s \cdot \frac{12d \cdot x_\gamma \log s}{r \cdot s^2} \right]$$

$$= r \left[z \cdot \frac{12d \cdot x_\gamma \log s}{r \cdot z^2} + \sum_{p=z+1}^{s} \frac{12d \cdot x_\gamma \log s}{rp^2} \right]$$

$$\leq \frac{24d \cdot x_\gamma \log s}{z}$$

packets with color γ are not marked. Summing over all colors yields the proposition. $\square$

Note that we showed above that for randomly distributed edges the m-distributor fails to fulfill the requirements of the balancing phase resp. the placement phase with probability at most $2/s$. Thus for randomly distributed edges a single m-distributor fails to fulfill the requirements of both the balancing phase *and* the placement phase with probability at most $4/s$. Hence for sufficiently large s there exists an m-distributor that can be used for both the balancing phase and the placement phase.

Next we analyze the delivery phase. Since each input node wants to send at most $\frac{s}{2k} \cdot k = \frac{s}{2}$ requests, the edges of the (s, m, k)-splitter can be distributed in such a way that each output node has degree $\frac{s}{2}$ and therefore receives at most $\frac{s}{2}$ requests. This ensures that, if an output node stores at most $\frac{rs}{2}$ packets at the beginning of the delivery phase, it can accept r new packets from every edge of the (s, m, k)-splitter without having more than $r \cdot s$ packets afterwards. Since output nodes with more than $\frac{rs}{2}$ packets may cause problems, we need a bound for the number of packets that can not be sent from the input nodes to the output nodes.

Proposition 8.1.5. *Let y be the total number of packets stored in the output nodes. For all $k \in \{1, \ldots, \sqrt{s}\}$ there exists an (s, m, k)-splitter for distributing the requests for the active packets in such a way among the outputs that at most $\frac{4y}{s}$ active packets fail to be sent to their output sets.*

Proof. Consider a fixed output set Y. Let y' be the number of packets stored in its nodes. The proof consists of a probabilistic argument. In particular, we show that for randomly chosen edges such that each input node has $\frac{s}{2k}$ and output node has $\frac{s}{2}$ endpoints the probability is less than one that, for any $y' \leq \frac{\varepsilon r s m}{k}$, there is a subset $C_{y'}$ of input nodes (of size depending on y') such that, for any choice of $\frac{s}{4k}$ edges out of the $\frac{s}{2k}$ edges leaving input nodes in $C_{y'}$, all remaining edges point to output nodes with more than $\frac{rs}{2}$ packets. Thus there exists a bipartite graph that restricts the number of active packets that fail to reach Y to be at most $\frac{r \cdot s}{4k} |C_{y'}|$.

Let m be the number of input and output nodes of the splitter and $c_{y'}$ be the size of $C_{y'}$. Then the probability described above is bounded by

$$\sum_{\substack{\frac{rs}{2} \leq y' \leq \frac{\varepsilon r s m}{k}, \\ \frac{rs}{2} | y'}} \binom{m}{c_{y'}} \left(\frac{\frac{s}{2k}}{\frac{s}{4k}}\right)^{c_{y'}} \binom{m/k}{\frac{y'}{rs/2}} \left(\frac{\frac{y'}{rs/2}}{m/k}\right)^{c_{y'} \cdot \frac{s}{4k}}$$

This is because the number of ways to choose $c_{y'}$ out of m input nodes for $C_{y'}$ is $\binom{m}{c_{y'}}$, and the number of ways to choose $\frac{s}{4k}$ out of the $\frac{s}{2k}$ edges leaving input nodes in $C_{y'}$ is bounded by $\binom{s/2k}{s/4k}^{c_{y'}}$. Furthermore there are at most $\frac{y'}{rs/2}$ output nodes with more than $\frac{rs}{2}$ packets. The number of ways to choose a subset of output nodes such that all output nodes with more than $\frac{rs}{2}$ packets are included is therefore bounded by $\binom{m/k}{y'/(rs/2)}$. The probability that one chosen edge from a node in $C_{y'}$ has its endpoint in such a subset is $\frac{y'/(rs/2)}{m/k}$. Since we require every output node to have a fixed degree of $\frac{s}{2}$, the probabilities for the $c_{y'} \cdot \frac{s}{4k}$ chosen edges to fall into the same subset of $y'/(rs/2)$ nodes are negatively correlated and therefore can be regarded as independent for an upper bound. Thus the probability that all of the $\frac{s}{4k}$ edges chosen for each input in $C_{y'}$ belong to a subset of output nodes with more than $\frac{rs}{2}$ packets is at most $(\frac{y'/(rs/2)}{m/k})^{c_{y'} \cdot s/4k}$. Let $c_{y'} \cdot \frac{s}{4k} = \frac{4y'}{rs}$ and s be sufficiently large (note that $k \leq \sqrt{s}$). Then the overall probability described above is bounded by

$$\sum_{\substack{\frac{rs}{2} \leq y' \leq \frac{\epsilon rsm}{k}, \\ \frac{rs}{2} \mid y'}} \binom{m}{c_{y'}} \left(\frac{s}{2k}\right)^{c_{y'}} \left(\frac{m/k}{\frac{s}{4k}}\right) \left(\frac{\frac{y'}{rs/2}}{\frac{y'}{rs/2}}\right) \left(\frac{\frac{y'}{rs/2}}{m/k}\right)^{c_{y'} \cdot \frac{s}{4k}}$$

$$= \sum_{z=1}^{2\epsilon m/k} \binom{m}{\frac{8k \cdot z}{s}} \left(\frac{s}{2k}\right)^{\frac{8k \cdot z}{s}} \left(\frac{m/k}{z}\right) \left(\frac{z}{m/k}\right)^{2z}$$

$$\leq \sum_{z=1}^{2\epsilon m/k} \left(\frac{es \cdot m}{8k \cdot z}\right)^{\frac{8k \cdot z}{s}} (2e)^{\frac{8k \cdot z}{s}} \left(\frac{em}{k \cdot z}\right)^{z} \left(\frac{z}{m/k}\right)^{2z}$$

$$\leq \sum_{z=1}^{2\epsilon m/k} \left(\frac{2s \cdot m}{k \cdot z}\right)^{\frac{8k \cdot z}{s}} \left(\frac{ek \cdot z}{m}\right)^{z}$$

$$\leq \sum_{z=1}^{2\epsilon m/k} \left(\frac{m}{ek \cdot z}\right)^{\frac{k \cdot z \log s}{s}} \left(\frac{ek \cdot z}{m}\right)^{z}$$

$$\leq \sum_{z=1}^{2\epsilon m/k} \left(\frac{ek \cdot z}{m}\right)^{z/2} \overset{\epsilon \leq \frac{1}{18e}}{\leq} \sum_{z=1}^{2\epsilon m/k} \left(\frac{1}{3}\right)^{z} < 1 .$$

Hence there exists an (s, m, k)-splitter such that $c_{y'} \leq \frac{4k}{s} \cdot \frac{4y'}{rs} = \frac{16ky'}{rs^2}$ for all $y' \leq \frac{\epsilon rsm}{k}$. Thus at most $\frac{r \cdot s}{4k} \cdot \frac{16ky'}{rs^2} = \frac{4y'}{s}$ active packets fail to be sent to Y. Summing over all output sets yields the proposition. $\qquad\square$

From Proposition 8.1.1, Proposition 8.1.3, and Proposition 8.1.5 it follows that the total number of packets stored in input nodes at the beginning of the phase reduces to at most

$$x' = \frac{48x \log s}{s} + \frac{384x \log s}{\sqrt{s}} + \frac{4y}{s} \leq \frac{400 \log s}{\sqrt{s}} (x + y)$$

packets at the end of the phase assuming that s is large enough. This completes the proof of Lemma 8.1.1. $\qquad\square$

Let X_i^t denote the number of packets in level i after the execution of phase t. Then the following theorem can be shown by using a potential function as described in [LM89] and [Up92].

Theorem 8.1.2. *There is a deterministic multi-port scheme on an r-replica-ted s-ary MBF of size N that routes any $r \cdot s$-relation from the top level to the bottom level in $O(\log_s N)$ steps if s is sufficiently large.*

Proof. We analyze the progress of the routing algorithm in terms of a potential function. Let n be the number of nodes in each level of a given (s, d, k)-MBF and $w = s^{1/4}/(\log s)^{1/2}$. The *potential* of a packet after phase t is w^i if after the execution of that phase the packet is in level $d - i$ of the network. (When a packet reaches its destination its potential is 1.) Let $\Phi(t)$ denote the sum of the potentials of the n packets after phase t, that is,

$$\Phi(t) = \sum_{i=0}^{d} X_i^t w^{d-i} \ .$$

Clearly $\Phi(0) = rs \cdot n \cdot w^d$, and routing a batch terminates at the first phase τ such that $\Phi(\tau) < r$.

Let $f(s) = \frac{400 \log s}{\sqrt{s}}$. Assume that t and i are even. Then by Lemma 8.1.1

$$X_i^{t+1} \leq f(s)(X_i^t + X_{i+1}^t) \quad \text{and} \quad X_{i+1}^{t+1} \leq X_i^t + X_{i+1}^t \ ,$$

and after the next phase

$$X_i^{t+2} \leq X_{i-2}^t + X_{i-1}^t + f(s)(X_i^t + X_{i+1}^t)$$

and

$$X_{i+1}^{t+2} \leq f(s)(X_i^t + X_{i+1}^t + f(s)(X_{i+2}^t + X_{i+3}^t)) \ .$$

Plugging these bounds into the potential function and applying the equation

$$X_{i+j}^t w^{d-i} = w^j \cdot X_{i+j}^t w^{d-(i+j)}$$

yields

$$\begin{aligned}
\Phi(t+2) \ &\leq \sum_{0 \leq i \leq d,\ i \text{ even}} (X_{i-2}^t + X_{i-1}^t + f(s)(X_i^t + X_{i+1}^t))w^{d-i} + \\
&\quad \sum_{0 \leq i \leq d,\ i \text{ odd}} f(s)(X_{i-1}^t + X_i^t + f(s)(X_{i+1}^t + X_{i+2}^t)))w^{d-i} \\
&\leq \sum_{0 \leq i \leq d,\ i \text{ even}} \left(\frac{1}{w^2} + f(s) + \frac{f(s)}{w} + f(s)^2 w \right) X_i^t w^{d-i} + \\
&\quad \sum_{0 \leq i \leq d,\ i \text{ odd}} \left(\frac{1}{w} + f(s)w + f(s) + (f(s)w)^2 \right) X_i^t w^{d-i} \\
&\leq \sum_{i=0}^{d} \left(\frac{1}{w} + f(s)w + f(s) + (f(s)w)^2 \right) X_i^t w^{d-i} \ .
\end{aligned}$$

Thus the potential function is decreased by at least

$$\frac{1}{w} + f(s)w + f(s) + (f(s)w)^2 = O(w^{-1})$$

every two phases, and for $\tau = O(\log_s N)$, $\Phi(\tau) < r$. Since there are only $O(1)$ batches to route the theorem follows.   □

We now show how to extend this scheme to route any global $r \cdot s$-relation in $O(\log_s N)$ steps. To simplify the presentation we use a topology with $3N$ nodes we simply call $G(r, s, d, k)$ for routing $r \cdot s \cdot N$ packets. $G(r, s, d, k)$ consists of $3d$ levels of $n = k\sqrt{s}^{d-1}$ nodes each and uses edges that can forward r packets in one step. The first d levels are connected by $(s, n, 1)$-routers, the second d levels represent the (r, s, d, k)-MBF, and the last d levels are connected to each other by $\frac{s}{2}$ forward edges between any input i and output i for all $i \in \{1, \ldots, n\}$.

Overlapping these three stages and identifying the corresponding nodes yields an N-node topology of depth d with degree $2(2s + s) + 2 \cdot \frac{s}{2} = 7s$, that can simulate one step in $G(r, s, d, k)$ with constant delay. Such a network is called *extended s-ary multibutterfly* and denoted by (r, s, d, k)-XBF.

Initially, all the $r \cdot s \cdot N$ packets reside in the first d levels of $G(r, s, d, k)$. All the final destinations of the packets are in the nodes of the last d levels. Clearly, there is a path with no more than $3d$ edges between every node in the first part and every node in the third part, and this path can be locally computed. A packet initially at node $\langle \ell, x \rangle$ with destination $\langle \ell', x' \rangle$ can take an arbitrary path forward to level d. By bit comparison, the packet is then led to the node $\langle 2d, x' \rangle$, and then by the direct edges $(\langle k, x' \rangle, \langle k+1, x' \rangle)$, the packet reaches its destination.

Each packet p with destination level q is assigned a fixed rank during the routing defined as $\mathrm{rank}(p) = q - 2d \in [d]$. For each node v in $G(r, s, d, k)$, we define the *median* of v to be the $\frac{r \cdot s}{2}$-largest rank in v if there are at least $\frac{r \cdot s}{2}$ packets in v and -1 otherwise.

8.1.4 Description of the Global Protocol

We partition the $rs \cdot N$ messages into 1566 batches such that no more than $rs \cdot m/1566$ messages from each batch that have the same destination level traverse any m-router in the MBF-levels. This can be done by declaring only those packets to be active at level j if their destination is in the set $\{z \mid z = j - 1 \pmod{1566}\}$. For the purpose of analysis we assume that all batch j packets have been routed before transmitting batch $j + 1$ packets, and we now concentrate on the routing of one batch.

Nodes at even levels of $G(r, s, d, k)$ are active in odd phases, nodes at odd levels are active in even phases. In one phase, the following routing strategies are performed in the three different parts of $G(r, s, d, k)$:

One Phase within the First d Levels. Consider an $(s, n, 1)$-router whose inputs are active. Let ρ denote the minimal rank such that the number of packets with rank $\geq \rho$ stored in the input and output nodes of that $(s, n, 1)$-router is at most $\frac{1}{522} rsn$. We perform the following two subphases.

The task of the *balancing phase* is to distribute the packets in such a way among the input nodes that there are only very few nodes with more than $\frac{r \cdot s}{4}$ packets with rank $\geq \rho$. Analogous to Proposition 8.1.1, this can be obtained if each input node sends out its $p \leq r \cdot s$ packets along the edges of the s-ary n-distributor with numbers 1 to $\lceil \frac{p}{r} \rceil$ such that packets with higher rank get edges with lower numbers.

The *delivery phase* consists of four steps. Its task is to approximately sort the packets according to their destination level (by moving packets with rank $\geq \rho$ forward and, maybe, in exchange packets backwards). For each input node, only the $\frac{r \cdot s}{4}$ packets with highest ranks are declared active. In the first step each input node sends a request message to $\frac{s}{2}$ suitably chosen output nodes. Each output node sends its median back to all input nodes that sent it a request. The input node then distributes its packets among the outgoing links such that packets with higher rank are preferred and (up to) r packets are sent along a link if their ranks are larger than the median. If the sum of the packets already stored at an output node and the packets it receives from the input nodes exceeds $r \cdot s$ then the output node sends in exchange to the new packets old packets back preferring packets with lower rank.

One Phase within the Second d Levels. Consider an (s, m, k)-router whose inputs are active. Let ρ be defined for the (s, m, k)-router as for the $(s, n, 1)$-router above. A phase consists of the following three subphases.

The task of the *balancing phase* is to distribute the packets in such a way among the input nodes that there are only very few nodes with more than $\frac{r \cdot s}{2}$ packets with rank $\geq \rho$. Analogous to Proposition 8.1.1, this can be obtained if each input node sends out its $p \leq r \cdot s$ packets along the edges of the s-ary n-distributor with numbers 1 to $\lceil \frac{p}{r} \rceil$ such that packets with higher rank get edges with lower numbers.

The task of the *placement phase* is to distribute the packets in such a way among the input nodes that there are only very few nodes having more than $\frac{r \cdot s}{4k}$ packets with rank $\geq \rho$ that have to be sent to the same output set. Analogous to Proposition 8.1.2 and 8.1.3, this can be obtained if all packets stored at input nodes are distributed among the s edges of the m-distributor with the help of a suitably chosen assignment graph preferring packets with higher ranks, before sending them out.

The *delivery phase* consists of four steps. Its task it to send as many packets as possible to output sets prescribed by their destinations. For each input node, only the first $\frac{r \cdot s}{4k}$ packets to each of the k output sets are declared active, preferring packets with higher rank. In the first step each input node sends a request message to $\frac{s}{2k}$ suitably chosen output nodes within each output set to which it has messages to transmit. Each output node sends its median back to all input nodes that sent it a request. The input node then distributes its packets among the outgoing links such that packets with higher rank are preferred and (up to) r packets are sent along a link if their ranks are larger than the median. If the sum of the packets already stored at an output node and the packets it receives from the input nodes exceeds $r \cdot s$ then the output node sends in exchange to the new packets old packets back preferring packets with lower rank.

One Phase within the Last d Levels. A phase simply consists of forwarding the packets along the $\frac{s}{2}$ edges for each active node.

8.1.5 Analysis of the Global Protocol

We first analyze the routing of one batch. Consider a subgraph connecting two levels in $G(r, s, d, k)$ in a phase in which the inputs of the subgraph are transmitting messages to the outputs. (For the first d levels this would be an $(s, n, 1)$-router, for the next d levels any (s, m, k)-router, and for the last d levels any two consecutive levels with active upper level.) Let $q \in [d]$ be fixed. Further let x_1 (resp. y_1) be the total number of packets with rank q or $q+1$ that are stored in the input nodes (resp. output nodes) at the beginning of that phase. Let x_2 (resp. y_2) be the number of packets with rank $\geq q+2$ stored in the input nodes (resp. output nodes) at the beginning of this phase. Moreover, let x' denote the total number of packets with rank q that are stored in the input nodes at the end of the phase. Then we can show the following lemma.

Lemma 8.1.2.

$$x' \leq \frac{400 \log s}{\sqrt{s}}(x_1 + y_1) + x_2 + y_2 \ .$$

Proof. The result trivially holds for the last d levels. Since the routing strategy in the first d levels is similar the strategy in the second d levels and makes use of $(s, n, 1)$-routers, it remains to prove the inequality above for any (s, m, k)-router in the second d levels, $k \in \{1, \ldots, \sqrt{s}\}$.

Let $\epsilon \leq \frac{1}{1566}$. We have to distinguish between two cases. If $x_2 + y_2 > \epsilon rsm$ then it immediately follows from the choice of the packets participating in one batch that $x' \leq x_2 + y_2$. Suppose in the following that $x_2 + y_2 \leq \epsilon rsm$. Then it holds that $x_1 + y_1 + x_2 + y_2 \leq 3\epsilon rsm \leq \frac{1}{522}rsm$. Since the packets with higher ranks are preferred in our protocol, we can apply Lemma 8.1.1 with $x = x_1 + x_2$ and $y = y_1 + y_2$ to the protocol above to get that $x' \leq \frac{400 \log s}{\sqrt{s}}(x_1 + y_1 + x_2 + y_2)$. Combining both cases yields the lemma.   □

Let $X_i^t(k)$ denote the number of packets with rank k in level i after the execution of phase t. Then the following theorem can be shown by using a potential function.

Theorem 8.1.3. *There is a deterministic multi-port scheme on an extended r-replicated s-ary MBF of size N that routes any global $r \cdot s$-relation in $O(\log_s N)$ steps if s is sufficiently large.*

Proof. We analyze the progress of the routing algorithm in terms of a potential function. Let $w = s^{1/4}/(\log s)^{1/2}$. The *potential* of a packet with rank k after phase t is w^i if after the execution of that phase the packet is in level $2d - i + k$ of the network. (When a packet reaches its destination its potential is 1.) Let $\Phi(t)$ denote the sum of the potentials of the n packets after phase t, that is,

$$\Phi(t) = \sum_{k=0}^{d-1} \sum_{i=0}^{2d+k} X_i^t(k) w^{2d-i+k} \ .$$

Clearly $\Phi(0) \leq rs \cdot N \cdot w^{3d}$, and routing a batch terminates at the first phase τ such that $\Phi(\tau) < r$.

Let $f(s) = \frac{400 \log s}{\sqrt{s}}$. Assume that t and i are even. Then by Lemma 8.1.2

$$X_i^{t+1}(k) \ \leq \ f(s)(X_i^t(k) + X_{i+1}^t(k) + X_i^t(k+1) + X_{i+1}^t(k+1)) +$$
$$\sum_{\ell=k+2}^{d-1} (X_i^t(\ell) + X_{i+1}^t(\ell))$$
and
$$X_{i+1}^{t+1}(k) \ \leq \ X_i^t(k) + X_{i+1}^t(k) \ .$$

And after the next phase

$$X_i^{t+2}(k) \ \leq \ X_{i-1}^{t+1}(k) + X_i^{t+1}(k)$$
$$\leq \ X_{i-2}^t(k) + X_{i-1}^t(k) + \tag{8.4}$$
$$f(s)(X_i^t(k) + X_{i+1}^t(k) + X_i^t(k+1) + X_{i+1}^t(k+1)) +$$
$$\sum_{\ell=k+2}^{d-1} (X_i^t(\ell) + X_{i+1}^t(\ell)),$$

and

$$X_{i+1}^{t+2}(k) \ \leq \ f(s)\left(X_{i+1}^{t+1}(k) + X_{i+2}^{t+1}(k) + X_{i+1}^{t+1}(k+1) + X_{i+2}^{t+1}(k+1)\right) + \tag{8.5}$$
$$\sum_{\ell=k+2}^{d-1} (X_{i+1}^{t+1}(\ell) + X_{i+2}^{t+1}(\ell))$$
$$\leq \ f(s)[\ X_i^t(k) + X_{i+1}^t(k) + \tag{8.6}$$
$$f(s)\left(X_{i+2}^t(k) + X_{i+3}^t(k) + X_{i+2}^t(k+1) + X_{i+3}^t(k+1)\right) +$$
$$\sum_{\ell=k+2}^{d-1} (X_{i+2}^t(\ell) + X_{i+3}^t(\ell)) +$$
$$X_i^t(k+1) + X_{i+1}^t(k+1) + f(s)\left(X_{i+2}^t(k+1)+\right.$$
$$X_{i+3}^t(k+1) + X_{i+2}^t(k+2) + X_{i+3}^t(k+2)) +$$
$$\sum_{\ell=k+3}^{d-1} (X_{i+2}^t(\ell) + X_{i+3}^t(\ell))\ ] + \sum_{\ell=k+2}^{d-1} [\ X_i^t(\ell) + X_{i+1}^t(\ell) +$$

$$f(s)(X_{i+2}^t(\ell) + X_{i+3}^t(\ell) + X_{i+2}^t(\ell+1) + X_{i+3}^t(\ell+1)) +$$

$$\sum_{\ell'=\ell+2}^{d-1} (X_{i+2}^t(\ell') + X_{i+3}^t(\ell'))\,].$$

Plugging these bounds into the potential function and applying the equation

$$X_{i+j}^t(k+\ell)w^{2d-i+k} = w^{j-\ell} \cdot X_{i+j}^t(k+\ell)w^{2d-(i+j)+(k+\ell)}$$

yields

$$\Phi(t+2) \leq \sum_{k=0}^{d-1}\left[\sum_{\substack{0\leq i\leq 2d+k,\\ i \text{ even}}} (8.4)\cdot w^{2d-i+k} + \sum_{\substack{0\leq i\leq 2d+k,\\ i \text{ odd}}} (8.6)\cdot w^{2d-i+k}\right]$$

$$\leq \sum_{k=0}^{d-1}\sum_{\substack{0\leq i\leq 2d+k,\\ i \text{ even}}}\left[\left(\frac{1}{w^2} + f(s)\left(1+\frac{1}{w}\right) + \sum_{j=2}^{d-1}\left(\frac{1}{w}\right)^j\right) + \right.$$

$$f(s)\left(1+\frac{1}{w}\right)\left(\frac{1}{w} + f(s)(w+1) + \sum_{j=1}^{d-1}\left(\frac{1}{w}\right)^j\right) +$$

$$\left.\sum_{j=0}^{d-1}\left(\frac{1}{w^{3+j}} + f(s)\left(\frac{1}{w^{1+j}} + \frac{1}{w^{2+j}}\right) + \sum_{\ell=3}^{d-1}\left(\frac{1}{w}\right)^{\ell+j}\right)\right].$$

$$X_i^t(k)w^{2d-i+k} +$$

$$\sum_{k=0}^{d-1}\sum_{\substack{0\leq i\leq 2d+k,\\ i \text{ odd}}}\left[\left(\frac{1}{w} + f(s)(w+1) + \sum_{j=1}^{d-1}\left(\frac{1}{w}\right)^j\right) + \right.$$

$$f(s)\left(1+\frac{1}{w}\right)\left(1 + f(s)(w^2+w) + \sum_{j=0}^{d-1}\left(\frac{1}{w}\right)^j\right) +$$

$$\left.\sum_{j=0}^{d-1}\left(\frac{1}{w^{2+j}} + f(s)\left(\frac{1}{w^j} + \frac{1}{w^{1+j}}\right) + \sum_{\ell=2}^{d-1}\left(\frac{1}{w}\right)^{\ell+j}\right)\right].$$

$$X_i^t(k)w^{2d-i+k}$$

$$\leq \sum_{k=0}^{d-1}\sum_{i=0}^{2d+k}\left[\left(\frac{1}{w} + f(s)(w+1) + \sum_{j=1}^{d-1}\left(\frac{1}{w}\right)^j\right) + \right.$$

$$f(s)\left(1+\frac{1}{w}\right)\left(1 + f(s)(w^2+w) + \sum_{j=0}^{d-1}\left(\frac{1}{w}\right)^j\right) +$$

$$\left.\sum_{j=0}^{d-1}\left(\frac{1}{w^{2+j}} + f(s)\left(\frac{1}{w^j} + \frac{1}{w^{1+j}}\right) + \sum_{\ell=2}^{d-1}\left(\frac{1}{w}\right)^{\ell+j}\right)\right].$$

$$X_i^t(k)w^{2d-i+k} \ .$$

Thus the potential function is decreased by at least

$$\left(\frac{1}{w} + f(s)(w+1) + \sum_{j=1}^{d-1} \left(\frac{1}{w}\right)^j \right) +$$

$$f(s)\left(1 + \frac{1}{w}\right)\left(1 + f(s)(w^2 + w) + \sum_{j=0}^{d-1} \left(\frac{1}{w}\right)^j\right) +$$

$$\sum_{j=0}^{d-1} \left(\frac{1}{w^{2+j}} + f(s)\left(\frac{1}{w^j} + \frac{1}{w^{1+j}}\right) + \sum_{\ell=2}^{d-1} \left(\frac{1}{w}\right)^{\ell+j} \right)$$

$$= \ O(w^{-1})$$

every two phases, and for $\tau = O(\log_s N)$, $\Phi(\tau) < r$. Since there are only $O(1)$ batches to route the theorem follows.   □

Combining the theorem with the results in [Up92] if s is not sufficiently large yields Theorem 8.1.1.   □

8.2 Universal Adaptive Routing Strategies

In the following we present three universal strategies for adaptive routing: greedy routing strategies, routing via sorting, and routing via simulation. We will show that the "routing via simulation" technique yields asymptotically optimal deterministic strategies for routing arbitrary permutations in any network with sufficiently large routing number.

8.2.1 Greedy Routing Strategies

One strategy to route packets deterministically in networks is to always try to get closer to the destination whenever possible. This strategy is usually called *greedy routing*.

Since greedy routing strategies are extremely difficult to analyze, only a few results are known so far in this area. Baran [Ba64] proposed the first greedy hot-potato algorithm. Since then, a lot of experimental results on greedy hot-potato routing have been published (see, e.g., [AS92, GG92, GH92]). Hajek [Ha91] presented a simple greedy hot-potato routing algorithm for the hypercube of size N that runs in $2k + N$ steps, where k is the number of packets. Ben-Aroya and Schuster [BS94] provided a greedy hot-potato routing algorithm for the $n \times n$ mesh that sends k packets with maximal source-to-destination distance D in $2(k-1) + D$ steps. (For an arbitrary permutation routing problem, this would mean a runtime of $O(n^2)$.)

Their result was generalized by Chinn *et al.* [CLT96]. They present a greedy algorithm for routing along shortest paths in an $n \times n$ mesh (a so-called *minimal* adaptive routing protocol) that routes any permutation in $O(n^2/k+n)$ time if each node of the mesh can buffer up to $k \geq 1$ packets. They further show that for a large class of minimal adaptive routing algorithms, the worst case time is bounded by $\Omega(n^2/k^2)$.

8.2.2 Routing via Sorting

In order to route packets deterministically in a network according to an arbitrary permutation, we can use the best known sorting algorithm for this network.

There are, for instance, routing algorithms based on AKS sorting [AKS83] that can be implemented on a constant degree network of size N with runtime $\Theta(\log N)$ [Le85]. Furthermore, the sorting algorithm developed by Cypher and Plaxton [CP93] implies that any permutation can be routed deterministically on a hypercube, shuffle-exchange, and cube-connected cycles of size N in time $O(\log N(\log \log N)^2)$. Kunde [Ku91] and Suel [Su94] showed that, for any constant d, there is a deterministic sorting algorithm for the d-dimensional mesh that is at most a factor of two away from the lower bound, which implies fast deterministic permutation routing protocols for these networks.

8.2.3 Routing via Simulation

In this section we present a very efficient deterministic routing protocol for arbitrary networks H. The idea of the protocol is that, for any permutation routing problem in H, we route the packets to their destinations by simulating routing in a suitably chosen extended s-ary multibutterfly embedded in H. For this we use the simulation strategy described in Section 6.3.2.

Description of the Protocol

Consider an arbitrary network H of size N with routing number R. In a preprocessing phase, we embed an extended R-ary multibutterfly of size approximately N/R in H such that every node in the multibutterfly is simulated by approximately R nodes in H. According to Lemma 8.2.1 such a multibutterfly exists.

Lemma 8.2.1. *For any $\sqrt{s}, n \geq 2$ there exist $k \in \{1, \ldots, \sqrt{s}\}$ and $d \geq 1$ such that the number n' of nodes in the (s, d, k)-XBF is bounded by $n/2 \leq n' \leq n$.*

Proof. According to the definition of the extended multibutterfly, the number of nodes in the (s, d, k)-XBF is $d \cdot k \cdot \sqrt{s}^{d-1}$. Choose $d \geq 1$ and $k \in \{1, \ldots, \sqrt{s}\}$

in such a way that the number n' of nodes gets maximal under the restriction that $n' \leq n$. We distinguish between two cases.

- $k < \sqrt{s}$: Then $d \cdot (k + 1) \cdot \sqrt{s}^{d-1} > n$ and therefore $d \cdot k \cdot \sqrt{s}^{d-1} \geq \frac{1}{2}d \cdot (k + 1) \cdot \sqrt{s}^{d-1} > n/2$.
- $k = \sqrt{s}$: Then $(d+1) \cdot 1 \cdot \sqrt{s}^d > n$ and therefore $d \cdot k \cdot \sqrt{s}^{d-1} \geq \frac{1}{2}(d+1) \cdot \sqrt{s}^d > n/2$.

This yields the lemma.   □

In the following, we assume that there exists an R-ary multibutterfly that has exactly N/R nodes. In this case we can assign exactly R nodes of H to any node of the multibutterfly. (If this is not possible, then Lemma 8.2.1 implies that an R-ary multibutterfly can be chosen such that we need clusters of size at most $2R$ to assign all nodes in H to nodes of the multibutterfly.) In order to partition the nodes of H into clusters of size R, we choose an arbitrary spanning tree T in H and apply the clustering strategy on T described in Section 6.3.2. Let the nodes of each cluster be connected by an Euler tour along edges in T. This ensures that

a) every Euler tour has a length of at most $6R$, and
b) the maximal number of Euler tours that share the same link is constant.

Furthermore we distribute the paths simulating edges of the multibutterfly among the nodes in H in such a way that every node is the endpoint of at most some constant number of paths. If we now want to route any permutation in H, we can transform this into the problem of routing any R-relation in the R-ary multibutterfly. Hence, in order to route a permutation in H, we can choose to perform a deterministic step-by-step simulation of routing an R-relation in the multibutterfly. With this strategy we can prove the following result (see also [MS96a]).

Theorem 8.2.1. *Let H be an arbitrary network of size N with routing number R. Then there is a deterministic online protocol that routes any permutation in time $O(\log_R N \cdot R)$, using only constant size buffers.*

Proof. Each step of the multibutterfly can be simulated in the following way in H.

- **Assigning XBF-edges to the packets:**
  In case that we have to simulate a balancing phase, this can be done by first sorting the packets in the nodes of each cluster according to their rank (note that the rank of a packet depends on its destination level in the XBF, see Section 8.1.4). Since only a constant number of Euler tours share the same link, this can be done in time $O(R)$ with constant size buffers, using Odd-Even Transposition Sort [Kn73]. Afterwards, for all $i \in \{1, \dots, R\}$ the packet with ith largest rank is sent along the Euler tour of its cluster until it reaches the node that simulates edge i of the R distributor edges leaving

that cluster. Clearly, this can also be coordinated among the clusters in time $O(R)$.

In case that we have to simulate a placement phase, we again first sort the packets in the nodes of each cluster according to their rank. As noted above, this takes $O(R)$ steps. Next the $R/2$ packets with highest ranks have to be assigned to suitably chosen distributor edges. For this, the nodes first count how many of these packets want to be sent to any of the at most $\sqrt{R}$ output sets. Using this information, the nodes compute which packet to forward along which distributor edge. (Note that this can be done by each node with the help of an algorithm presented by Vazirani [Va94] that runs in $O(R^{3/2})$ time to find an assignment of packets to edges. Since a calculation step can be usually performed much faster than a communication step, we will not consider the time for computing such an assignment.) This information is used to distribute the packets among the nodes of the cluster. Clearly, the counting and distribution of packets can be performed in $O(R)$ steps for every cluster, using only constant size buffers.

In case that we have to simulate the first step of a delivery phase, we generate a request packet for each node simulating the endpoint of a splitter edge. After all answers of the requests are received (for this we need the routing strategy below), the packets in each cluster are first sorted according to the output set they want to reach and their rank, and then delivered among the nodes that received a positive answer, preferring packets with higher rank. As above, the sorting and distribution of the packets can be performed in $O(R)$ steps for every cluster, using only constant size buffers.

– **Moving each packet along its assigned XBF-edge:**
According to the definition of the routing number, the edges of the XBF can be simulated by a path collection in H with congestion $O(R)$ and dilation at most R. Since at most one packet is sent along each of these paths, there is an offline protocol according to Theorem 6.2.1 that routes packets along these paths in time $O(R)$ using only constant size edge buffers.

Combining these results with Theorem 8.1.1 yields the time bound of the theorem. □

Applications

Theorem 8.2.1 has the following applications.

Corollary 8.2.1. *For any network H of size N with routing number $R = \Omega(N^\epsilon)$ for some constant $\epsilon > 0$ there exists a deterministic routing strategy that routes any permutation in H in $O(R)$ steps, using constant size buffers.*

$R = \Omega(N^\epsilon)$ holds, for instance, for all networks with diameter $\Omega(N^\epsilon)$ or bisection width $O(N^{1-\epsilon})$. According to [SV93], every n-vertex graph of genus g and maximal degree d has bisection width $O(\sqrt{gdn})$. Thus the following result is true.

Corollary 8.2.2. *For any network H of size N with genus g and degree d such that $g \cdot d = O(N^{1-\epsilon})$, $\epsilon > 0$ constant, there exists a deterministic routing strategy that routes any permutation in H in $O(R)$ steps, using only constant size buffers.*

This result implies somewhat surprisingly that for *any planar* network with degree $O(N^{1-\epsilon})$ there is an asymptotically optimal deterministic routing strategy.

8.3 Summary of Main Results

At the end of this chapter we summarize its main results. We saw in previous chapters that, if random bits are available, oblivious routing protocols together with Valiant's trick already have the potential of reaching the routing number for arbitrary networks. However, if no random bits are available, adaptive strategies are needed. First, we considered deterministic routing in specific networks. In particular, we were able to show the following result (see Theorem 8.1.1).

Given an extended r-replicated s-ary multibutterfly of size N with $r \geq 1$ and $s \geq 2$, $r \cdot s \cdot N$ packets, $r \cdot s$ per processor, can be routed deterministically according to some arbitrary $r \cdot s$-relation in time $O(\log_s N)$.

In order to get efficient deterministic routing protocols for arbitrary networks, we chose a technique called "routing via simulation": Given a network H of size N with routing number R, any permutation routing problem for it can be routed efficiently in a deterministic way by embedding an R-ary multibutterfly of size N/R in it and interpreting the permutation routing problem for H as an R-relation routing problem for the multibutterfly. The outcome was the following result (see Theorem 8.2.1).

Let H be an arbitrary network of size N with routing number R. Then there is a deterministic online protocol that routes any permutation in time $O(\log_R N \cdot R)$, using only constant size buffers.

This theorem implies that for *any* planar graph of size N with degree at most $O(N^{1-\epsilon})$ for some constant $\epsilon > 0$ there is an asymptotically optimal deterministic permutation routing strategy that only needs constant size buffers.

9. Compact Routing Protocols

In this chapter we present results in the field of compact routing. The first section gives an overview about what has been done so far in the area of compact routing. In the following sections we deal with upper bounds for the relationship between space and slowdown that hold for any network. We start with describing in Section 9.2 a universal compact routing strategy called "routing via simulation". This technique is used in Section 9.3 and Section 9.4 to develop efficient randomized and deterministic compact routing protocols. The efficiency will in both cases be measured using the routing number. Afterwards we show how to apply these protocols to specific networks.

9.1 History of Compact Routing

In the following we give an overview on the results that have been presented in the two areas of compact routing dealing with the relationship between space and stretch factor, and the relationship between space and slowdown.

9.1.1 Relationship between Space and Stretch Factor

The issue of saving space in routing tables by settling for near-shortest path routings was first raised by Kleinrock and Kamoun [KK77]. In [KK77], clustering approaches for networks of general topology are studied.

Most previous work focused on solutions for special classes of network topologies. Optimal (stretch factor 1) routing schemes with total memory requirement $O(N \log N)$ were designed for simple topologies like hypercubes [DS87], acyclic graphs [KS85], unit-cost rings, complete networks and grids [LT86, LT87], and outerplanar graphs [FJ88].

In [FJ89], Frederickson and Janardan present two routing schemes for planar graphs. The first scheme achieves a stretch factor of 3 using $O(\log N)$-bit names and a total of $O(N^{4/3} \log N)$ bits for storing routing information in the nodes. For any constant ϵ, $0 < \epsilon < \frac{1}{3}$, the second scheme guarantees stretch factor 7, has a total memory requirement of $O((1/\epsilon)N^{1+\epsilon} \log N)$ bits, and uses $O((1/\epsilon) \log N)$-bit names. Both schemes are separator-based, the first using the separator strategy of [LT79], and the second the more structured cyclic separator of [Mi86].

In [FJ90] Frederickson and Janardan present two space-efficient near-shortest path routing schemes for any class of networks whose members can be decomposed recursively by a separator of cutsize at most some constant c, so-called c-decomposable networks. For any such network of size N, the first scheme has stretch factor at most 3 and uses a total of $O(cN \log^2 N)$ bits of routing information, and $O(\log N)$-bit names, generated from a separator based decomposition of the network. The second scheme augments the node names with $O(c \log c \log N)$ additional bits which results in a total memory requirement of $O(c^2 \log c \cdot N \log^2 N)$ bits, and uses this to reduce the stretch factor to $1 + 2/a$ where $a > 1$ is the positive root of the equation $a^{\lceil (c+1)/2 \rceil} - a - 2 = 0$.

Peleg and Upfal [PU89b] show that any routing scheme for general N-node networks that achieves a stretch factor $k \geq 1$ must use a total of $\Omega(N^{1+1/(2k+4)})$ bits of routing information in the nodes. Further they present a family of hierarchical routing schemes for unit-cost general networks, which guarantees a stretch factor of $12k + 3$, $O(\log^2 N)$-bit labels for the nodes, $O(\log N)$-bit headers, and a total space requirement of $O(k^3 N^{1+1/k} \log N)$. In the special case of chordal graphs G, they show that a stretch factor of at most 3 can be obtained if a total amount of $O(N \log^2 N)$ bits is available for storing routing information.

Awerbuch et al. [ABLP90] present two families of hierarchical routing schemes that improve the upper bound in [PU89b]. The first scheme guarantees a stretch factor $2^k - 1$ and requires storing a total of $O(k \cdot N^{1+1/k} \log^2 N)$ bits. The second scheme guarantees a stretch factor of $2 \cdot 3^k - 1$ and requires at most $O(k \log N \cdot (d + N^{1/k}))$ bits for storing routing information in a node of degree d, and $O(kN^{1+1/k} \log N)$ bits overall. They also describe an efficient distributed preprocessing algorithm for this scheme.

In [AP92], Awerbuch and Peleg present a routing scheme that allows non-uniform cost on the edges. Given a graph with diameter D, it guarantees a stretch factor of at most $16k^2$, requiring $O(k \cdot N^{1/k} \log N \cdot \log D)$ bits per node for storing routing information. Headers, and node labeling are of size $O(\log N)$ bits.

Fraigniaud and Gaviolle [FG94] show that for all unit circular-arc graphs of size n and degree d, a stretch factor of 1 can be obtained if $O(d \log N)$ bits are available in each node for storing routing information.

In [FG96], Fraigniaud and Gaviolle show that for any stretch factor < 2 there exist networks of size N in which $\Omega(N^\epsilon)$ nodes require $\Omega(N \log N)$ bits for storing routing information, $\epsilon > 0$ constant.

Table 9.1 summarizes the best known upper bounds for the local and global memory requirement, and the best known bounds on the memory complexity of universal routing schemes on networks of size N as a function of the stretch factor s. When no reference is indicated, the complete routing table is the best known routing scheme.

Table 9.1: Best known bounds on the memory requirement.

stretch	local memory requirement	global memory requirement
$s = 1$	$\Theta(N \log N)$ [GP95]	$\Theta(N^2 \log N)$ [GP95]
$1 < s < 2$	$\Theta(N \log N)$ [FG96]	$\Omega(N^2)$ [FG95] $O(N^2 \log N)$
$2 \leq s < 3$	$\Omega(N^{1/(2s+4)})$ [PU89b] $O(N \log N)$	$\Omega(N^{1+1/(2s+4)})$ [PU89b] $O(N^2 \log N)$
$3 \leq s < 16$	$\Omega(N^{1/(2s+4)})$ [PU89b] $O(N \log N)$	$\Omega(N^{1+1/(2s+4)})$ [PU89b] $O(N^{1+1/\lfloor \log(s+1) \rfloor} \log^2 N)$ [ABLP90]
$16 < s \leq 87$	$\Omega(N^{1/(2s+4)})$ [PU89b] $O(N^{1/\lfloor \sqrt{s}/4 \rfloor} \log^2 N)$ [AP92]	$\Omega(N^{1+1/(2s+4)})$ [PU89b] $O(N^{1+1/\lfloor \log(s+1) \rfloor} \log^2 N)$ [ABLP90]
$s \geq 87$	$\Omega(N^{1/(2s+4)})$ [PU89b] $O(\sqrt{s} \cdot N^{1/\lfloor \sqrt{s}/4 \rfloor} \log^2 N)$ [AP92]	$\Omega(N^{1+1/(2s+4)})$ [PU89b] $O(s^3 N^{1+1/\lfloor (s-3)/12 \rfloor} \log N)$ [PU89b]
$s = O(\log N)$	$\Omega(1)$ $O(e^{\sqrt{\log N}} \log^{3/2} N)$ [AP92]	$\Omega(N)$ [PU89b] $O(N \log^4 N)$ [PU89b]
$s = O(\sqrt{N})$	$\Omega(1)$ $O(e^{\sqrt{\log N}} \log^{3/2} N)$ [AP92]	$\Omega(N)$ [PU89b] $O(N \log N)$ [PU89b]

Table 9.1 shows that for $s < 2$ there are nodes that need $\Omega(N \log N)$ bits, but for $s \geq 16$ there are schemes that can do better. Therefore an important open question is, where in between 2 and 16 we have for the first time the situation that all nodes need $o(N \log N)$ bits.

The protocols presented above usually use one (or a combination) of the following two strategies.

Interval Routing. Let G be a network of size N. In the ILS (interval labeling scheme), node labels belong to the set $[N]$, assumed cyclically ordered, and each link is assigned an interval in $[N]$ such that for each node its adjacent links have disjoint intervals, and the union of the intervals is $[N]$. To transmit a message m from node u to w, m is sent by u along the (unique) link $e = (u, v)$ whose interval contains the label of w. With this approach, one always obtains an optimal memory usage, while the problem is to choose labels for the nodes and intervals for the links in such a way that messages are routed along shortest paths. The ILS can be generalized to the k-ILS, where k intervals are assigned to each link. Interval schemes have been analyzed, e.g., in [LT86, LT87, FJ88, FGS93, FG94].

Hierarchical Routing. Hierarchical routing strategies partition the nodes of a network into clusters that are themselves partitioned into subclusters, and so on. Each cluster usually has a node we call *center* that knows how to get to the centers of next higher or next lower clusters. The further the distance a packet has to be sent, the higher upwards it has to be routed along the hierarchy of clusters by visiting its centers. The problem is to find for a given stretch factor s a hierarchical clustering such that messages can be sent from any source to any destination along a path with stretch factor s. Hierarchical routing schemes have been presented, e.g., in [PU89b, ABLP90].

9.1.2 Relationship between Space and Slowdown

In case that a network is only lightly loaded with messages, the stretch factor is a very accurate parameter for measuring the quality of path systems. This, however, is only true for lightly loaded networks, since the stretch factor does not give any information about how many paths share the same link. So for highly loaded networks the congestion can be catastrophic even for path systems with constant stretch factor. Therefore we will use another approach for the rest of this chapter that considers both the dilation and the expected congestion of path systems. These path systems will be used to design routing strategies that have a very low slowdown even under severe space restrictions. In this area, only a few results have been published so far.

In case of interval routing, nothing general is known about the relationship between the congestion and dilation of the resulting path system and the routing number of the underlying network. It has been shown, however, that there exist n-node networks with diameter D that require an $\Omega(\sqrt{n})$-ILS to have a path system with paths of length at most $1.5D - 3$ [KRS96].

All hierarchical schemes have the great disadvantage that the routing is done with the help of a clustering of the graph, where some vertices are declared as routing centers for a set of other vertices. Hence these strategies cause a very high congestion if randomly chosen functions have to be routed. Therefore hierarchical routing schemes are not useful to obtain a small congestion and therefore a fast routing time.

In [MS95a] Meyer auf der Heide and Scheideler were the first who proved a nontrivial upper bound for the trade-off between the space for storing routing information and the slowdown of routing in arbitrary node-symmetric networks. In particular, they showed that for any node-symmetric network of size N with degree d and diameter $D = \Omega(\log N)$ it holds for every $s \in \{2, \ldots, N\}$ that any permutation can be routed in $O(\log_s N \cdot D)$ steps (that is, with slowdown $O(\log_s N)$), w.h.p., if $O(s \cdot D \cdot \log d + \log N)$ space is available at each node and $O(\log(s \cdot D) + \log \log N)$ space is available in each packet for storing routing information. This result was further improved by Meyer auf der Heide and Scheideler in [MS96a]. It is shown there that for any network of size N with degree d and routing number R, and any $s \in \{\log N, \ldots, R\}$ if $R \geq \log N$ and $s = R$ otherwise, any permutation can be routed deterministically in time $O(\log_s N \cdot R)$ (and therefore slowdown $O(\log_s N)$) if $O(s \cdot d \log d)$ space is available at every node and $O(\log(s \cdot R))$ space is available in every packet for storing routing information.

The strategy they use to achieve these results is called the "routing via simulation" strategy. In the following we describe how this strategy works when applied to compact routing.

9.2 The "Routing via Simulation" Strategy

Consider the problem of storing routing information in the nodes of a network in such a way that routing an arbitrary permutation can be done time- and space-efficiently. Our approach to achieve a relationship between the slow-down and space requirements for storing routing information in a network is to use the "routing via simulation" technique:

Consider $G = (W, F)$ to be the guest graph and $H = (V, E)$ the host graph with $m := \frac{|V|}{|W|} \geq 1$. Then each node u in G is represented by m nodes called *copies* of u in H, each simulating to be the endpoint of a subset of edges leaving u in G. Let $\mathcal{P}_C$ be a path collection connecting for each node u in G its m copies in H. Furthermore let $\mathcal{P}_F$ be a path collection which contains paths $p_H(u, v)$ between nodes u and v in H only if u and v simulate the endpoints of an edge $e \in F$. Our strategy to simulate routing in G by H then works as follows:

Suppose, a packet with origin u and destination v travels along the path $p_G(u, v)$ in G. In order to simulate the traversal of an edge $\{w, w'\} \in p_G(u, v)$, it first uses a path in $\mathcal{P}_C$ to get to the node in H simulating the starting point of $\{w, w'\}$ and then uses a path in $\mathcal{P}_F$ to get to the node simulating the endpoint of $\{w, w'\}$.

To keep the development of programs for a parallel system independent from the techniques to obtain space-efficient routing, we restrict ourselves to number the nodes in H consecutively from 0 to $N - 1$. For the simulation of a graph G with n nodes by H we therefore need a function h that maps $[N]$ to $[n]$ telling node $i \in [N]$ in H that packets stored in its buffers belong to node $h(i) \in [n]$ in G. In the ideal case, h can be defined as $h(x) = \lfloor \frac{x}{m} \rfloor$ (that is, N is a multiple of n). This can be done by assigning the node in H that simulates the ith copy of node $x \in [n]$ in G the number $m \cdot x + i$.

We will describe later how a function h that only needs space $O(\log N)$ can also be developed for the general case (that is, N is not a multiple of n). In the following sections we present strategies to choose suitable path collections in H for $\mathcal{P}_C$ and $\mathcal{P}_F$, and design space-efficient methods for storing routing information.

9.2.1 Selecting Suitable Routing Structures

In this section we present a strategy to select routing structures in H for our simulation strategy that have low congestion and dilation.

Consider the copies of the nodes in G to be partitioned into subsets of size $\Theta(c)$ for some fixed c. Then we apply the strategy described in Section 6.3.2 to an arbitrary spanning tree T in H in order to partition the nodes of H into clusters of size $\Theta(c)$ such that each subset of copies is simulated by a cluster of the same size. Let the nodes of each cluster be connected by an Euler tour along edges in T. Then the following two results hold (see Section 6.3.2 for a proof)

a) the Euler tour of every cluster has length at most $O(c)$, and
b) the maximal number of Euler tours that share the same link is constant.

Hence we can connect subsets of copies via cycles in H that have a low congestion. (In order to ensure, that they also have a small dilation, c, of course, should not be too large.) Furthermore we may need paths in H for connecting copies in different clusters. In order to bound the congestion and dilation of these paths, we will use the following lemma.

Lemma 9.2.1. *Consider any network H with routing number R. Let H be partitioned into clusters of size c. Assume that each of these clusters has to be connected to at most d other clusters. Then there exists a simple path collection in H for establishing these connections with dilation at most R and congestion $O(\frac{d \cdot R}{c} + \log(R + \frac{c}{d}))$.*

Proof. We distinguish between two cases. If $d \geq c$ then choose each node in H to be the endpoint of at most $\lceil d/c \rceil$ connections. Thus the problem of establishing a path for each connection reduces to the problem of finding an efficient path collection for an arbitrary $\lceil d/c \rceil$-relation routing problem. Because of the definition of R, there exists a simple path collection for any such problem with congestion at most $R \cdot \lceil d/c \rceil$ and dilation at most R.

Consider now the case that $d < c$. We will show with the help of the Lovász Local Lemma that there exists a path collection, one path for each connection, with congestion $O(\frac{d \cdot R}{c} + \log(R + \frac{c}{d}))$ and dilation at most R.

For any connection between any cluster C_1 and C_2, $\lfloor c/d \rfloor$ pairs $\{u, v\}$ of nodes are chosen as *candidates*, $u \in C_1$, $v \in C_2$. These candidates can be chosen such that each node belongs to at most one candidate. Hence there exists a collection of simple paths, one for each candidate, with congestion at most R and dilation at most R. Now consider the random experiment of choosing randomly for each connection one of the $\lfloor c/d \rfloor$ paths representing its candidates and eliminating the rest. We associate a bad event to each edge e in H. The bad event for edge e is that more than k surviving paths contain e (k will be determined later). To show that there is a way of choosing the candidates such that no bad event occurs, we need to bound the dependence b among the bad events and the probability p of each individual bad event occurring.

The dependence calculation is straightforward. Whether or not a bad event occurs depends solely on the selection of the candidates that pass through the corresponding edge. Since at most R candidates pass through an edge, and each of these candidates belongs to a set of $\lfloor c/d \rfloor$ candidates for a connection, each having a length of at most R, the dependence b of the bad events is at most $R \cdot \lfloor c/d \rfloor \cdot R$.

Next we compute the probability of each bad event. Let p be the probability of the bad event corresponding to edge e. Then

$$p \leq \binom{R}{k} \left(\frac{1}{\lfloor c/d \rfloor}\right)^k \leq \left(\frac{e \cdot R}{k \cdot \lfloor c/d \rfloor}\right)^k .$$

For $k \geq \max\{\frac{2e \cdot R}{\lceil c/d \rceil}, 3 \log(R + \frac{c}{d})\}$ the product $ep(b + 1)$ is less than 1, and thus, by the Lovász Local Lemma, there is a choice of candidates such that the congestion is $O(\frac{d \cdot R}{c} + \log(R + \frac{c}{d}))$.   □

9.2.2 Space-Efficient Perfect Hashing

In this section we present basic definitions in the field of hashing. As we will see in the next section, hashing can be used to construct space-efficient data structures for storing routing information in the nodes of H.

Let the data set S comprise n elements belonging to the universe $U = \{0, 1, \ldots, m - 1\}$. Further let M be an arbitrary nonempty set. Consider the problem of storing a function $f : S \to M$ in a hash table T. Each entry of T can store an element $x \in S$ and $f(x)$. A sequence $H = (h_1, \ldots, h_k)$ of functions is a *perfect k-probe hash function* for S, if $H : U \to [1, n]^k$, and $T[1 \cdots n]$ can be organized so that each item $x \in S$ is located in one of the k probe positions defined by applying the k-probe functions to x. Then for $k > 1$ the value of $f(x)$ for any $x \in S$ can simply be obtained by calculating $h_i(x)$ for all $i \in \{1, \ldots, k\}$ and testing whether x is stored in $T(h_i(x))$. For $k = 1$ we only require T to store f. In [SS90], Schmidt and Siegel present the following theorem.

Theorem 9.2.1. *An $O(1)$-time perfect 1-probe hash function for a set of n elements belonging to the universe $[m]$ can be specified by $O(n + \log \log m)$ bits, which matches the lower bound to within a constant factor.*

9.2.3 Design of Compact Routing Tables

In this section we present a method to design compact routing tables for storing $\mathcal{P}_C$ and $\mathcal{P}_F$ in the nodes in H. First we consider the problem of connecting different clusters via paths in H.

Lemma 9.2.2. *Let H be a network with degree d and $\mathcal{P}$ be a path collection in H with dilation D and congestion C. Further let P be an upper bound for the number of paths in $\mathcal{P}$ that have their endpoint at a common node in H. Then H needs at most $O(C \cdot d \log d + P \log(d \cdot C \cdot D))$ space in each vertex for storing $\mathcal{P}$.*

Proof. According to the definition of C and D, every path in $\mathcal{P}$ shares its links with at most $C \cdot D$ other paths in $\mathcal{P}$. Suppose $G' = (V', E')$ is a graph in which each node represents a path in $\mathcal{P}$ and nodes $x, y \in V'$ are connected with each other if their respective paths share a link in H. Then G' has a degree of at most $d' = C \cdot D$. According to Theorem 3.3.2, $d' + 1$ colors suffice to color every d'-regular graph in such a way that no two adjacent vertices have the same color. Therefore it is possible to attach numbers to the paths in $\mathcal{P}$ out of $[C \cdot D + 1]$ in such a way that no two paths in $\mathcal{P}$ with a common link have the same number.

Let $\psi : \mathcal{P} \to [C \cdot D + 1]$ be the function that assigns a number to all paths in $\mathcal{P}$ such that the condition above is fulfilled. Then we choose the following strategy to store routing information.

Consider any node v in H. Let E_v be the set of all edges in H with endpoint v and $\mathcal{P}_v$ be the set of all paths in $\mathcal{P}$ that have their endpoint at v. In order to store $\mathcal{P}$ we need the following two tables.

- $T_{v,1} : \mathcal{P}_v \to E_v \times [C \cdot D + 1], p \to (e, \psi(p))$ maps each path in $\mathcal{P}_v$ to a suitable color and the first edge e used by this path in H.
- $T_{v,2} : E_v \times [C \cdot D + 1] \to E_v \cup \{\emptyset\}$ is arranged such that $T_{v,2}(e, k)$ contains the edge the path with number k entering v via e uses to leave v.

Clearly, it takes at most $O(P \log(d \cdot C \cdot D))$ bits to store $T_{v,1}$ and $O(d \cdot C \cdot D \cdot \log d)$ bits to store $T_{v,2}$. If we apply perfect hashing techniques (see Theorem 9.2.1) we can reduce the size of $T_{v,2}$ to $O(d \cdot C \log d)$ bits in such a way that we can still evaluate $T_{v,2}$ in constant time by a hash function that needs $O(d \cdot C + \log \log(d \cdot C \cdot D))$ bits. Altogether this results in routing tables of size

$$O(C \cdot d \log d + P \log(d \cdot C \cdot D))$$

per node for storing $\mathcal{P}$. $\square$

Consider now the problem of routing packets within a cluster of size $\Theta(c)$. Together with the results in Section 9.2.1 we can show the following lemma.

Lemma 9.2.3. *Let d be the degree of H. Further let the nodes of H be partitioned into clusters of size $\Theta(c)$ as described in Section 9.2.1. Then each node needs at most $O(d \log d + \log c)$ space to store all Euler tours traversing it.*

Proof. Let $\mathcal{E}$ be the set of all Euler tours in H. According to Section 9.2.1, every Euler tour shares its edges with at most $C = \Theta(c)$ other Euler tours. Then $C + 1$ colors suffice to color the Euler tours in such a way that no two Euler tours with the same color share an edge. Thus we can choose the following strategy to store routing information.

Consider any node v in H. Let E_v be the set of all edges in H with endpoint v. In order to store $\mathcal{E}$ we have to store its number k in v, and the edge the Euler tour uses to leave v. Further we need the following table.

- $T_{v,3} : E_v \times [C + 1] \to E_v$ is arranged such that $T_{v,3}(e, k)$ contains the next edge the Euler tour with number k entering v via e uses to leave v.

Clearly, it takes at most $O(d \cdot C \cdot \log d)$ bits to store $T_{v,3}$. Since the Euler tours have constant congestion, we can apply perfect hashing techniques (see Theorem 9.2.1) to reduce the size of $T_{v,3}$ to $O(d \log d)$ bits in such a way that we can still evaluate $T_{v,3}$ in constant time by a hash function that needs $O(d + \log \log(d \cdot C))$ bits. Moreover, v needs $O(\log C)$ bits to store the color of the cluster it belongs to. $\square$

Note that we need additional space in the nodes of H for storing its number, $\mathcal{P}_G$ and h. This depends on the choice of G and the different simulation techniques we will describe later. Our strategy will be to choose G in such a way that the space requirements for the path collections connecting the clusters dominate the space necessary to store the other structures.

The tables described in Lemma 9.2.2 and Lemma 9.2.3 can be used in the following way. Suppose, we want to send a packet p along a path P in $\mathcal{P}_C$ or $\mathcal{P}_F$. Let p be currently stored at the endpoint v of P in H. Suppose that p knows the number of the other endpoint in H. (This is the case if the nodes know the topology of G and the way how the endpoints of edges adjacent to a node in G are distributed among its copies in H.) We have to distinguish between two cases.

If P is a path connecting two different clusters then we use the tables described in Lemma 9.2.2. First, p gets the color c and the first edge e of the path P by accessing $T_{v,1}$. The packet chooses to traverse e and stores the color c in its routing information. Let e' be the last edge p used so far to reach some node u. With the help of $T_{u,2}$, e' and its actual color, the packet determines the edge it has to traverse next in H. p continues to access $T_{u,2}$ for each node u it visits until it reaches the other endpoint of P (in this case, we have $T_{u,2}(e',c) = \emptyset$).

If P is a path connecting two nodes within one cluster, we use the table described in Lemma 9.2.3. First, v provides p with the color c of the cluster it belongs to, and the next edge e of the Euler tour p has follow in that cluster. The packet chooses to traverse e and stores the color c in its routing information. Let e' be the last edge p used so far, and p be currently stored in node u. With the help of $T_{u,3}$, e' and its actual color, the packet determines the edge it has to traverse next along the Euler tour. It continues to access $T_{u,3}$ for each node u it visits until it reaches the other endpoint of P.

9.3 Randomized Compact Routing

In this section we present a randomized compact routing protocol. Since the s-ary butterfly network has a very regular and therefore space-efficient structure, we will use as guest graph G a variant of the s-ary butterfly which is defined as follows.

9.3.1 The (s, d, k)-Butterfly

For $k \in \{1, \ldots, s\}$, the (s, d, k)-BF consists of a node set

$$V = \{(l, x) \mid l \in [d+1],\ x = (x_{d-1}, \ldots, x_0) \in [k] \times [s]^{d-1}\}$$

and can be derived from k $(s, d-1)$-BFs as shown in Figure 9.1 (for a definition of the $(s, d-1)$-BF see Section 4.2.1).

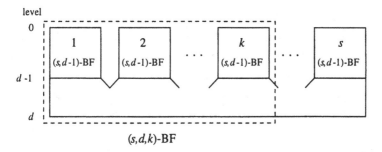

Fig. 9.1. An (s, d, k)-BF in an (s, d)-BF.

The (s, d, k)-WBF is defined by taking the (s, d, k)-BF and identifying level d with level 0. The next lemma will be important for our proofs. Its proof is similar to that of Lemma 8.2.1.

Lemma 9.3.1. *For any $s, n \geq 2$ there exist $k \in \{1, \ldots, s\}$ and $d \geq 1$ such that the number n' of nodes in the (s, d, k)-WBF is bounded by $n/2 \leq n' \leq n$.*

Given any $s, n \geq 2$, let $G(s, n)$ denote in the following the (s, d, k)-WBF whose size n' is closest to n.

9.3.2 The Simulation Strategy

Consider any network H with N nodes and routing number R. In order to describe how to embed an s-ary butterfly in H, we have to distinguish between the cases $2s \geq R$ and $\log R \leq 2s < R$.

The Simulation Strategy for $2s \geq R$. Consider the graph $G(s, N/2)$. We first modify $G(s, N/2)$ such that each node has two *copies* as described in Figure 9.2. The resulting graph is called $G'(s, N/2)$.

Fig. 9.2. Splitting the nodes into two copies.

Let $R_s, R'_s = \Theta(R)$ be chosen such that $(\lfloor \frac{2s}{R} \rfloor - 1)R_s + R'_s = 2s$. Then each s-clique in this modified graph can be partitioned into $(\lfloor \frac{2s}{R} \rfloor - 1)$ sets of

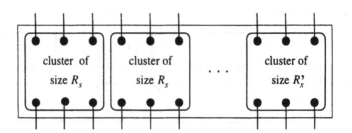

Fig. 9.3. Partitioning the nodes of an s-clique into clusters.

R_s nodes and one set of R'_s nodes as described in Figure 9.3. (By "s-clique" we mean here a complete bipartite graph consisting of two sets of s nodes.)

In case that $k < s$ in the (s, d, k)-WBF representing $G(s, N/2)$, we choose one of the following strategies for k-cliques connecting level $d - 1$ to level 0 in $G'(s, N/2)$.

- $2k \geq R$: In this case we use the same strategy as described for s above with cluster sizes R_k and R'_k.
- $2k < R$: Let n be the number of nodes in a level of $G'(s, N/2)$. Then we choose $R_k, R'_k = \Theta(R)$ in such a way that R_k and R'_k are multiples of $2k$, and $(\lfloor \frac{2n}{R} \rfloor - 1)R_k + R'_k = 2n$. The k-cliques are partitioned into $(\lfloor \frac{2n}{R} \rfloor - 1)$ clusters consisting of $R_k/2k$ k-cliques, and one cluster consisting of $R'_k/2k$ k-cliques.

According to Section 9.2.1, we can cluster the nodes in H in such a way that each cluster described above can be embedded 1-1 into a cluster of the same size in H whose nodes can be connected via a cycle of length at most six times the size of that cluster. If the size of $G'(s, N/2)$ is less than N then we choose larger cluster sizes in H for the clusters in G. According to Lemma 9.3.1, these cluster sizes have to be at most twice as large as the cluster sizes for the case that the size of $G'(s, N/2)$ equals N. Hence, each node in $G'(s, N/2)$ can be simulated by at most 2 nodes in H. As can be seen in the proof of Lemma 9.3.9, this does not change the runtime of our simulation strategy asymptotically. Therefore, in order to simplify the rest of the proof, we only describe the simulation strategy for the case that the size of $G'(s, N/2)$ is equal to N. First we show how to establish a path collection for the simulation of one step of $G'(s, N/2)$.

Let the path collection $\mathcal{P}_C$ be responsible for connecting for each node in $G(s, N/2)$ its two copies in H, and $\mathcal{P}_F$ be responsible for simulating the edges of $G(s, N/2)$ in H. $\mathcal{P}_F$ consists of the following three path collections.

- $\mathcal{P}_F^i$ contains a path for every pair of copies that lie in the same cluster in $G'(s, N/2)$,
- $\mathcal{P}_F^s$ contains R paths for every pair of clusters that lie in a common s-clique in $G'(s, N/2)$, and

– $\mathcal{P}_F^k$ contains R paths for every pair of clusters that lie in a common k-clique connecting level $d - 1$ to level 0 of $G'(s, N/2)$. (Note that for $2k < R$ $\mathcal{P}_F^k$ is empty.)

If we require the endpoints of the paths for each cluster to be distributed evenly among its nodes, we get the following results.

Lemma 9.3.2. $\mathcal{P}_C$ *can be embedded in H with congestion at most R and dilation at most R.*

Proof. Since the construction of $\mathcal{P}_C$ can be transformed to the problem of finding a path collection in H for routing some fixed partial permutation, the lemma follows directly from the definition of the routing number.   □

Lemma 9.3.3. $\mathcal{P}_F^i$ *can be embedded in H along edges in T with congestion $O(1)$ and dilation $O(R)$.*

Proof. For each cluster, $\mathcal{P}_F^i$ simply contains an Euler tour along the edges in T that connects all nodes in that cluster. As described above, each cluster contains $\Theta(R)$ nodes. Hence it follows from Section 9.2.1 that the congestion of the Euler tours is $O(1)$ and the dilation $O(R)$.   □

Lemma 9.3.4. $\mathcal{P}_F^s$ *can be embedded in H with congestion $O(s)$ and dilation at most R.*

Proof. The number of paths in $\mathcal{P}_F^s$ leaving each cluster is at most $d = O(R \cdot \frac{s}{R})$. Furthermore, each cluster consists of $c = \Theta(R)$ nodes. Applying Lemma 9.2.1 with these values for c and d yields the lemma.   □

Analogous to Lemma 9.3.4, the following lemma holds.

Lemma 9.3.5. $\mathcal{P}_F^k$ *can be embedded in H with congestion $O(k)$ and dilation at most R.*

The Simulation Strategy for $\log R \leq 2s < R$. In case that $\log R \leq 2s < R$, we embed $G(s, \epsilon N)$ in H with $\epsilon = 1/(2\lfloor \frac{R}{2s} \rfloor)$. For this, we first modify $G(s, \epsilon N)$ in a way that each node gets $2\lfloor \frac{R}{2s} \rfloor$ copies, $\lfloor \frac{R}{2s} \rfloor$ for each s-clique, as shown in Figure 9.4. The resulting graph is called $G''(s, \epsilon N)$.

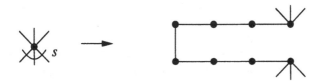

Fig. 9.4. Splitting the nodes into $2\lfloor \frac{R}{2s} \rfloor$ copies.

Since each s-clique in $G(s, \epsilon N)$ has $2s \cdot \lfloor \frac{R}{2s} \rfloor \approx R$ copies, we represent it as one cluster. The k-cliques connecting level $d-1$ to level 0 in $G(s, \epsilon N)$ are combined to clusters, each consisting of approximately $\lfloor \frac{s}{k} \rfloor$ such subgraphs, in a similar way as described for $2k < R$ for the case $2s \geq R$.

According to Section 9.2.1 we can cluster the nodes in H in such a way that each cluster described above can be embedded 1–1 into a cluster of the same size in H whose nodes can be connected via a cycle of length at most six times the size of that cluster. If the size of $G'(s, \epsilon N)$ is less then N, then we slightly increase the cluster sizes as described for the case $2s \geq R$. In the following we only describe how to simulate routing for the case that the size of $G'(s, \epsilon N)$ is equal to N.

Let the path collection $\mathcal{P}_C$ be responsible for connecting the copies of nodes in H, and let $\mathcal{P}_F$ be responsible for simulating the edges of $G(s, \epsilon N)$ in H. $\mathcal{P}_C$ consists of the following two path collections.

- $\mathcal{P}_C^i$ contains a path for any pair of copies of the same node in the same cluster in $G'(s, \epsilon N)$, and
- for every node in $G(s, \epsilon N)$, $\mathcal{P}_C^e$ contains a path that connects its two sets of copies in different clusters in $G'(s, \epsilon N)$.

Then we get the following results.

Lemma 9.3.6. $\mathcal{P}_C^i$ and $\mathcal{P}_F$ can be embedded in H along edges in T with congestion $O(1)$ and dilation $O(R)$.

Proof. Analogous to the proof of Lemma 9.3.3, we represent $\mathcal{P}_C^i$ and $\mathcal{P}_F$ by Euler tours along edges in T. □

Lemma 9.3.7. $\mathcal{P}_C^e$ can be embedded in H with congestion $O(s)$ and dilation at most R.

Proof. Since each cluster contains copies of $\Theta(s)$ nodes in $G(s, \epsilon N)$, $\mathcal{P}_C^e$ has $d = \Theta(s)$ paths leading out of each cluster. The size of all clusters is bounded by $c = \Theta(R)$. Applying Lemma 9.2.1 with these values for c and d yields the lemma. □

9.3.3 Bounding the Congestion and Dilation

Consider the problem of routing an arbitrary permutation in H using the path collections described above. If we use Valiant's trick we can reduce this to the problem of routing random functions in H. Since we want to route packets in H by simulating the routing in some suitable network $G(s, n)$, we are interested in the routing performance of $G(s, n)$ for randomly chosen relations (if each node in $G(s, n)$ is simulated by at most m nodes in H, then routing a random function in H can be transformed into the problem of simulating the routing of a random m-relation in $G(s, n)$). Analogous to the proof of Lemma 7.5.5 the following properties can be shown for $G(s, n)$.

Lemma 9.3.8. *There is a randomized online strategy for $G(s,n)$ using paths of dilation at most $2\log_s n$ such that, for a randomly chosen function, the expected congestion at*

- *any node in $G(s,n)$ is at most $2\log_s n$,*
- *any edge in an s-clique in $G(s,n)$ is at most $\frac{2\log_s n}{s}$,*
- *any edge in a k-clique connecting level $d-1$ to level 0 in $G(s,n)$ is at most $\frac{2\log_s n}{k}$.*

Let S denote the routing strategy for $G(s,n)$ described in Lemma 7.5.5. During the simulation of $G(s,n)$ by H, let us call a packet to be at *superstage* q if it is currently sent along the path (or paths) simulating the qth edge on its way to its destination in $G(s,n)$ using S. Since S sends packets along paths in $G(s,n)$ of length at most $2\log_s N$ (note that $n \le N/2$), and every edge in $G(s,n)$ is simulated by at most two stages of simple paths in H (one for routing within a cluster, and one for routing between different clusters), the overall number of stages used by the packets during the simulation is at most $4\log_s N$. It remains to bound the *stage congestion* that holds in H w.h.p. Note that the stage congestion is defined as the maximum over all stages of the maximal number of packets in a fixed stage that want to use the same edge in H during the simulation of $G(s,n)$ (see also Section 7.5.2). Lemma 9.3.8 can be used to prove the following lemma.

Lemma 9.3.9. *Applying strategy S to routing a randomly chosen function in H yields a stage congestion of $O(R + \log N)$, w.h.p., and a stage dilation of $O(R)$.*

Proof. The bound for the stage dilation follows from Lemma 9.3.2 to Lemma 9.3.7. In order to bound the stage congestion, we need the following definitions.

Let the *expected edge contention* be defined as the maximum over all edges e in $G(s,n)$ and superstages q of the expected number of packets at superstage q that traverse edge e during the routing of a randomly chosen function. Further let the *expected node contention* be defined as the maximum over all nodes of the sum of the expected edge contention of all edges adjacent to it.

From Lemma 9.3.8 we know that the expected edge congestion caused by strategy S in $G(s,n)$ is at most $\frac{2\log_s n}{s}$. Since S only allows the packets to be sent downwards in $G(s,n)$, we get that, if at most m packets with random destinations start in each node in $G(s,n)$, then the expected edge contention in every s-clique in $G(s,n)$ is at most $\frac{m}{\log_s n} \cdot \frac{2\log_s n}{s} = \frac{2m}{s}$. Similarly, the expected edge contention in the edges connecting level $d-1$ to level 0 is at most $\frac{2m}{k}$. In order to bound the stage congestion in H, we have to distinguish between two cases.

$2s \ge R$: Since each node of $G(s,n)$ gets two copies in H, H has to simulate the routing of a randomly chosen 2-relation in $G(s,n)$. Hence we get for

- $\mathcal{P}_C$: Since the congestion of $\mathcal{P}_C$ is at most R, and the expected node contention in $G(s, n)$ is at most 4, the expected stage congestion is at most $4R$.
- $\mathcal{P}_F^i$: Since $\mathcal{P}_F^i$ has congestion $O(1)$ and dilation $O(R)$, and the expected node contention in $G(s, n)$ is at most 4, it follows that the expected stage congestion is $O(R)$.
- $\mathcal{P}_F^s$, $\mathcal{P}_F^k$: Each cluster belonging to a subgraph $B(s)$ has R paths to each of the other clusters in $B(s)$. Since each path has to simulate $\Theta(R)$ edges in $G(s, n)$, and the expected contention of any edge in $B(s)$ is $\frac{4}{s}$, the expected stage contention for each of these paths is at most $O(\frac{R}{s})$. Together with Lemma 9.3.4 this yields an expected stage congestion of $O(R)$. The same can be shown for clusters that belong to one or contain several subgraphs $B(k)$.

For any fixed stage q, let the random variable X_e^q denote the number of packets at stage q that traverse edge e in H during the simulation of routing in $G(s, n)$ by H. Further let, for every packet p, the binary random variable $X_{p,e}^q$ be one if and only if p traverses e at stage q during the simulation. Then $X_e^q = \sum_p X_{p,e}^q$. Since the probabilities for the $X_{p,e}^q$ are independent and $E[X_e^q] = O(R)$, we can apply Chernoff bounds to get that the stage congestion for e is bounded by $O(R + \log N)$, w.h.p. Thus the stage congestion for the whole simulation of routing a random 2-relation in $G(s, n)$ is bounded by $O(R + \log N)$, w.h.p.

$\log R \leq 2s < R$: In this case H has to simulate the routing of a randomly chosen $2\lfloor \frac{R}{2s} \rfloor$-relation in $G(s, n)$. Hence we get for

- $\mathcal{P}_C^i$ and $\mathcal{P}_F$: According to Lemma 9.3.6, both path collections have congestion $O(1)$ and dilation $O(R)$. Since the expected node contention in $G(s, n)$ is at most $4\lfloor \frac{R}{2s} \rfloor$, and each node in $G(s, n)$ is represented by $2\lfloor \frac{R}{2s} \rfloor$ copies in H, we get an expected stage congestion of $O(R)$.
- $\mathcal{P}_C^e$: Each cluster has $\Theta(s)$ paths to other clusters in H, one path for each node in $G(s, n)$. Since the expected node contention in $G(s, n)$ is $4\lfloor \frac{R}{2s} \rfloor$, and the congestion of $\mathcal{P}_C^e$ is $O(s)$, we get an expected stage congestion of $O(R)$.

Analogous to the case $2s \geq R$ it follows that the stage congestion is also bounded by $O(R + \log N)$ w.h.p. for the case $\log R \leq 2s < R$. $\square$

Thus we can prove the following theorem.

Theorem 9.3.1. *Let H be an arbitrary network with N nodes, degree d and routing number $R = \Omega(\log N)$. Then, for every $s \in \{\log R, \ldots, N\}$, there is a randomized path selection scheme for routing any permutation along $O(\log_s N)$ stages of simple path collections with stage congestion $O(R)$, w.h.p., and stage dilation $O(R)$, if $O(s \cdot d \log d + \log N)$ space is available at* *each node and $O(\log(s \cdot R))$ space is available in each packet for storing routing information.*

Proof. The bounds on the stage congestion and stage dilation follow from Lemma 9.3.9. It remains to prove the space necessary to store routing information in the nodes and packets.

First we bound the space necessary to store the paths that connect different clusters. According to Lemma 9.3.2, 9.3.4, 9.3.5, and 9.3.7, the congestion of these path collections is at most $O(s)$. Furthermore the maximal number of these paths that have their endpoint at a common node in H is at most $O(\max\{1, \frac{s}{R}\})$. Thus according to Lemma 9.2.2 the space necessary in each node of H to store the path collections connecting clusters is bounded by $O(s \cdot d \log d + \max\{1, \frac{s}{R}\} \log(R \cdot d)) = O(s \cdot d \log d)$. The paths within a cluster simply follow a prescribed Euler tour. According to Lemma 9.2.3, this needs space $O(d \log d + \log R)$ in each node. For every node, the space necessary for storing its identification number and h is bounded by $O(\log N)$ (for h, it suffices to store R_s, R_s', R_k, R_k', s, k, and d). Combining the results yields the space bound in Theorem 9.3.1 for the nodes. The packets have to store the color of the path they are currently using. This needs $O(\log(s \cdot R))$ bits. $\qquad\square$

9.3.4 Applications

Theorem 9.3.1 together with the routing protocol by Ostrovsky and Rabani (see Theorem 7.9.1) for routing packets within a stage yields the following theorem.

Theorem 9.3.2. *Let H be an arbitrary network with N nodes, degree d and routing number $R = \Omega(\log^{1+\epsilon} N)$ for some constant $\epsilon > 0$. Then, for every $s \in \{\log R, \ldots, N\}$, there is a randomized online protocol that routes an arbitrary permutation in H in time $O(\log_s N \cdot R)$ (i.e., slowdown $O(\log_s N)$), w.h.p., if $O(s \cdot d \log d + \log N)$ space is available at each node and $O(\log(s \cdot R))$ space is available in each packet for storing routing information.*

Proof. The runtime bound immediately follows from Theorem 9.3.1 and Theorem 7.9.1. It therefore remains to bound the space requirements. The protocol by Ostrovsky and Rabani requires the packets to store information about their current track, level and delay, which requires $O(\log \log n)$ bits. The space required for the nodes to store the protocol by Ostrovsky and Rabani is bounded by $O(\log N)$, since in essence all it does is to check whether congestion or contention bounds of some size at most $poly(\log N)$ are violated, and assigning new delays to packets if necessary, or stopping packets when they managed a certain frame of size at most $poly(\log N)$ in order to synchronize with others. For more details see Section 2 in [OR97]. $\qquad\square$

If we restrict ourselves to node-symmetric networks, we can replace the simple path collections by shortcut-free path collections. This enables us to use the extended growing rank protocol which yields the following result.

Theorem 9.3.3. *Let H be an any node-symmetric network with N nodes, degree d and diameter $D = \Omega(\log N)$. Then, for every $s \in \{\log D, \ldots, N\}$, there is a randomized online protocol that routes an arbitrary permutation in H in time $O(\log_s N \cdot R)$ (i.e., slowdown $O(\log_s N)$), w.h.p., if $O(s \cdot d \log d + \log N)$ space is available at each node and $O(\log(s \cdot D))$ space is available in each packet for storing routing information.*

Proof. According to Theorem 5.1.2 and Theorem 5.2.2, there exists a concatenation of two shortest path collections that can connect the nodes in H according to an arbitrary permutation with congestion $O(D)$ and dilation at most $2D$. Using this in the construction of the path collections above that connect different clusters, yields the congestion, dilation, and space bounds in Theorem 9.3.1 with $R = D$. Hence, according to Theorem 7.5.1, the extended growing rank protocol applied to our construction yields a routing time of $O(\log_s N \cdot D)$ steps, w.h.p., which implies a slowdown of $O(\log_s N)$.

It remains to consider the space requirements for the extended growing rank protocol. Since the range of the ranks within one stage is bounded by $O(D)$ and there are $O(\log_s N)$ stages, each node only needs $O(\log D + \log \log_s N)$ bits to store the protocol. Besides the color of its path, each packet has to store its rank. According to Theorem 7.5.1, this takes $O(\log D)$ bits. Combining the space bounds concludes the proof of the theorem.   □

Note that the space bounds for the nodes in both theorems do not consider the space needed for storing packets. The protocol for Theorem 9.3.2 requires buffers of size $poly(\log N)$, whereas the protocol for Theorem 9.3.3 requires buffers of size $O(D \log_s N)$.

It follows, e.g., from Theorem 9.3.3 that for all bounded degree node-symmetric networks of size N with diameter D we get: If only space $O(\log N)$ is allowed for each vertex (which is optimal) and space $O(\log D)$ is allowed for storing routing information in a packet, the routing of an arbitrary permutation finishes after $O(\frac{\log N}{\log \log N} D)$ steps, w.h.p. If a space of $O(N^\epsilon)$, $\epsilon > 0$ constant, is allowed for each vertex and space $O(\log N)$ is allowed for each packet to store routing information, the routing finishes after $O(D)$ steps, w.h.p.

9.4 Deterministic Compact Routing

In case that we want to design a deterministic compact routing protocol we use as guest graph G the extended r-replicated s-ary multibutterfly defined in Section 8.1. The idea is to simulate each routing step in G with the help of an offline protocol in H. In order to bound the space requirements for this offline protocol we need the following lemma.

Lemma 9.4.1. *Consider an arbitrary simple path collection with congestion C and dilation D such that the sources of the paths are disjoint and all nodes have degree at most d. Suppose that along each path p packets have to be sent. Then there exists an offline protocol that can route all packets in time $O(p \cdot C + D)$ if constant size buffers are available at each edge and $\Theta(d \cdot C + \log p)$ space is available at each node for storing the protocol.*

Proof. Let the packets be divided into p batches such that each batch contains exactly one packet for every path.

Consider first the case that $C \geq D$. Then we can use the offline protocol described in Theorem 6.2.1 to route any batch in time $O(C + D) = O(C)$, using only constant size edge buffers. Hence altogether we need time $O(p \cdot C)$ to route all packets. Since every batch uses the same collection of paths, we can use the same offline protocol for every batch. We therefore only need to store in the nodes the offline protocol for one batch and a counter for the number of batches that have already been routed. Clearly, the counter needs $O(\log p)$ space. Every packet waits at most $O(C)$ time steps in its source before it traverses its first edge. This needs $O(\log C)$ space, because we assume that the sources of the paths are disjoint. Since each packet only has to wait a constant number of steps in any buffer once it has started (see [LMR94]), each edge needs at most $O(C)$ space to coordinate arriving packets by using a table with entries for every time point of the protocol. An entry is 0 if no packet arrives at this time and otherwise contains the number of steps a packet arriving at this time point has to wait. As every node has degree at most d, $\Theta(d \cdot C + \log p)$ space suffices in each node to execute the offline protocol for all batches.

Consider now the case that $C < D$. Let all paths be divided into subpaths of length at least C and at most $2C$. (If a path has length below C then it is considered as one single subpath.) Let a subpath be called *intermediate* if it is neither the first nor the last subpath of the path it belongs to. We want to choose an edge in each intermediate subpath such that no edge is chosen more than once. Let us call edges with this property *secure*. In order to show that a secure edge can be assigned to every intermediate subpath for any choice of the subpaths obeying the length constraints above in any path collection, consider the following construction.

Let $G = (V_1, V_2, E)$ be a bipartite graph with V_1 representing the intermediate subpaths and V_2 representing all edges used by the path collection. A node $u \in V_1$ is connected to node $v \in V_2$ if the subpath representing u contains the edge representing v. Since each intermediate subpath has length at least C, all nodes in V_1 have degree at least C. Furthermore, every node in V_2 has degree at most C, because every edge is used by at most C paths and therefore by at most C subpaths. Hence it holds for every subset $U \subseteq V_1$ that

$|\Gamma(U)| \geq |U|$. Otherwise there must exist a node in $\Gamma(U)$ with degree at least $C+1$. From Theorem 3.3.1 it follows that there must exist a matching of size $|V_1|$ in G. Thus for each intermediate path a secure edge can be chosen.

For every path, let its first edge be the secure edge of its first subpath. Consider now the situation that each secure edge has one packet in its buffer, and every packet has to be sent to the next secure edge (or the destination) on its respective path. This routing problem has congestion $O(C)$ and dilation $O(C)$. Hence the offline protocol described in Theorem 6.2.1 can be used to route all packets in time $O(C)$, using only constant size edge buffers at any time of the execution.

Our goal now is to interpret the secure edges as intermediate destinations, and to send the batches of packets along these intermediate destinations in a pipelined fashion using the offline protocol above, starting with batch 1-packets followed by batch 2-packets, and so on. If we use this strategy to route the p batches of packets along their respective paths, the runtime and requirements for the buffer size of the offline protocol above imply that the overall runtime is bounded by $O(p \cdot C + D)$, using only constant size edge buffers. Analogous to the case $C \geq D$ each node needs $O(d \cdot C + \log p)$ space to execute the offline protocol for all batches.   □

In the following we describe how to use this offline strategy.

9.4.1 The Simulation Strategy

Let H be an arbitrary network with N nodes and routing number R. Given any $s, n \geq 2$, let $G(r, s, n)$ denote the (r, s, d, k)-XBF whose size n' is closest to n. According to Lemma 8.2.1 it holds that $n/2 \leq n' \leq n$.

Let $s \leq R$. We partition the nodes of H into $\lfloor N/R \rfloor$ clusters of size R using the strategy described in Section 9.2.1, each simulating a single node in $G(\lfloor \frac{R}{s} \rfloor, s, \frac{N}{R})$. If the size of $G(\lfloor \frac{R}{s} \rfloor, s, \frac{N}{R})$ multiplied by R is less than N then some nodes in H are not assigned to clusters. In this case we only have to increase the size of the clusters by a factor of at most two. Hence we can show the following lemma.

Lemma 9.4.2. *For any network H of size N with routing number R and clusters of size $\Theta(R)$, one for each node of $G(\lfloor \frac{R}{s} \rfloor, s, \frac{N}{R})$, $s \in \{\log R, \ldots, R\}$, there is a simple path collection for simulating the edges in $G(\lfloor \frac{R}{s} \rfloor, s, \frac{N}{R})$ with congestion $O(s)$.*

Proof. Since each node of $G(\lfloor \frac{R}{s} \rfloor, s, \frac{N}{R})$ has degree $O(s)$ and each cluster in H has $\Theta(R)$ nodes, Lemma 9.2.1 yields that there is a path collection that simulates the edges in $G(\lfloor \frac{R}{s} \rfloor, s, \frac{N}{R})$ with congestion $O(s)$.   □

With this result we can show the following theorem.

Theorem 9.4.1. *Let H be an arbitrary network with N nodes, degree d, and routing number R. Then, for every $s \in \{\log R, \ldots, R\}$, there is a deterministic online protocol that routes any permutation in time $O(\log_s N \cdot R)$ (i.e., with slowdown $O(\log_s N)$), if constant size buffers are available at each edge for storing packets, $O(s(d \log d + \log s) + \log N)$ space is available at each node, and $O(\log(s \cdot R))$ space is available at each packet for storing routing information.*

Proof. Since each edge in $G(\lfloor \frac{R}{s} \rfloor, s, \frac{N}{R})$ has $\lfloor \frac{R}{s} \rfloor$ channels, and each node in $G(\lfloor \frac{R}{s} \rfloor, s, \frac{N}{R})$ is simulated by $\Theta(R)$ nodes in H, the channels can be distributed among the nodes such that every node is assigned to at most a constant number of channels. Then each step of the multibutterfly can be simulated in the following way in H.

- **Assigning XBF-channels to the packets:**
  For this we basically use the same strategies as described in Theorem 8.2.1, with the difference that here we require the packets to be distributed among nodes simulating endpoints of channels instead of endpoints of edges.
- **Moving each packet along its assigned XBF-channel:**
  In order to route the packets along their assigned XBF-channel, consider the packets to be separated into batches, each batch representing a different channel number. Since the Euler tours have constant congestion, it is easy to send the packets batch after batch to the starting point of the path they have to take if for each XBF-edge the nodes simulating its channels are ordered from channel 1 to channel $\lfloor \frac{R}{s} \rfloor$ along their Euler tour and the node simulating channel 1 represents the starting point of the path simulating that XBF-edge.
  As shown above, the edges of the XBF can be simulated by a path collection in H that has dilation at most R and congestion $O(s)$. Since at most $\frac{R}{s}$ packets are sent along each of these paths, the congestion of the path collection is bounded by $O(R)$. Hence we can use the strategy described in Lemma 9.4.1 to route the packets in batches along the paths in time $O(R)$, using only constant size buffers.

Combining these results with Theorem 8.1.1 yields the time bound of the theorem. It remains to prove an upper bound for the space necessary to store routing information in the nodes and packets. Let d be the degree of H.

- **Storing the embedding of $G(\lfloor \frac{R}{s} \rfloor, s, \frac{n}{R})$ into H:**
  Similar to the proof of Theorem 9.3.1, we need $O(\log N)$ bits to store a function h in each node telling it that a node with number x simulates node $h(x)$ in $G(\lfloor \frac{R}{s} \rfloor, s, \frac{N}{R})$.
- **Storing the Euler tours:**
  According to Lemma 9.2.3, $O(d \log d + \log R)$ bits suffice to store a lookup table in such a way that for each cluster the packets can be routed along an Euler tour.

- **Storing the paths simulating edges in the XBF:**
  Since at most $O(s)$ paths traverse any edge in H, we can use techniques described in Lemma 9.2.2 to show that $O(s \cdot d \log d)$ bits suffice to store a lookup table for the at most $O(s \cdot d)$ paths crossing each node.
- **Storing routing information in the packets:**
  According to Section 9.2.3, packets have to be able to store the color of the path they are currently following. Since at most $O(s)$ paths cross any link in H, and the dilation of the path collection is at most R, this can be done according to Lemma 9.2.2 by using $O(\log(s \cdot R))$ bits. Note that the rank of a packet can be computed with the help of its destination address and h and therefore does not need any bits in the packets.
- **Storing the XBF structure:**
  Consider a node v in H that is the endpoint of a path simulating an edge in the XBF. If this edge belongs to a distributor, v needs its number, which takes $O(\log s)$ bits. If this edge belongs to a splitter, v needs to know to which output set it leads. This also takes $O(\log s)$ bits. Hence each node requires $O(\log s)$ bits to store the XBF structure.
- **Storing the offline protocol:**
  According to Lemma 9.4.1, constant size buffers are sufficient to route the packets in batches along their paths. Since the congestion of the path collection is bounded by $O(s)$, we furthermore need at most $O(d \cdot s)$ bits in each node to store the offline protocol.
- **Storing information about the assignment of paths:**
  Since the placement phase is the only phase that requires working with a special assignment graph, it suffices to consider the space requirements for simulating the placement phase.
  According to Proposition 8.1.2, $O(s\sqrt{s}\log s)$ space is necessary to store the assignment graph of an s-ary XBF. This can be distributed among the nodes of a cluster such that each node needs at most $O(\sqrt{s}\log s)$ space for storing a part of that assignment graph. Furthermore, each node needs $O(\sqrt{s}\log s)$ space to store the number of packets in its cluster that have to be sent to any of the $\sqrt{s}$ output sets. Given a fixed number of packets to each of these output sets, each node that stores nodes of the assignment graph representing one block of size $\lfloor \frac{R}{s} \rfloor$ of these packets sends out packets containing information about edges adjacent to these nodes. Since nodes in the assignment graph representing blocks of packets have constant degree, the number of bits necessary to store information about edges adjacent to any such node of the assignment graph is at most $O(\log s)$. Hence at most $O(s)$ packets have to be sent along the Euler tour to inform all nodes of the respective cluster about the subgraph of the assignment graph for which a maximum matching has to be found. Storing this subgraph requires $O(s \log s)$ bits in each of the nodes. Using Vazirani's algorithm [Va94], it further takes $O(s \log s)$ space to compute a maximum matching.

Combining all space requirements yields the space bound above.   □

9.4.2 Applications

Theorem 9.4.1 has the following implications.

Corollary 9.4.1. *For any bounded degree network of size N with routing number $R = \Omega(N^\epsilon)$, $\epsilon > 0$ constant, $O(N^\delta)$ space suffices in the nodes for any constant $\delta > 0$ to obtain a routing time of $O(R)$ (i.e., a constant slowdown) for any permutation.*

Corollary 9.4.2. *For any bounded degree network of size N with routing number R, $O(\log R \cdot \log \log R + \log N)$ space suffices in the nodes to route any permutation in time $O(\frac{\log N}{\log \log N} R)$ (i.e., with slowdown $O(\frac{\log N}{\log \log N}))$.*

9.5 Summary of Main Results

Let us conclude this chapter with a summary of its main results. We presented a technique called "routing via simulation" that can be used to construct compact routing protocols for arbitrary networks. First, we constructed a randomized compact routing protocol for arbitrary node-symmetric networks. In particular, we were able to show the following results (see Theorems 9.3.2 and 9.3.3).

Let H be an arbitrary network with N nodes, degree d and routing number $R = \Omega(\log^{1+\epsilon} N)$ for some constant $\epsilon > 0$. Then, for every $s \in \{\log R, \ldots, N\}$, there is a randomized online protocol that routes an arbitrary permutation in H in time $O(\log_s N \cdot R)$ (i.e., slowdown $O(\log_s N))$, w.h.p., if $O(s \cdot d \log d + \log N)$ space is available at each node and $O(\log(s \cdot R))$ space is available in each packet for storing routing information.

Let H be an any node-symmetric network with N nodes, degree d and diameter $D = \Omega(\log N)$. Then, for every $s \in \{\log D, \ldots, N\}$, there is a randomized online protocol that routes an arbitrary permutation in H in time $O(\log_s N \cdot R)$ (i.e., slowdown $O(\log_s N))$, w.h.p., if $O(s \cdot d \log d + \log N)$ space is available at each node and $O(\log(s \cdot D))$ space is available in each packet for storing routing information.

The first result implies that for all bounded degree networks of size N with routing number $R = \Omega(\log^{1+\epsilon} N)$ for some constant $\epsilon > 0$ we get: If only space $O(\log N)$ is allowed for each vertex (which is optimal) and space $O(\log R)$ is allowed for storing routing information in a packet, the routing of an arbitrary permutation finishes after $O(\frac{\log N}{\log \log N} R)$ steps, w.h.p. If a space of $O(N^\epsilon)$, $\epsilon > 0$ constant, is allowed for each vertex and space $O(\log N)$ is allowed for each packet to store routing information, the routing finishes after $O(R)$ steps, w.h.p.

We also developed deterministic compact routing protocols. In particular, we could show the following result (see Theorem 9.4.1).

Let H be an arbitrary network with N nodes, degree d, and routing number R. Then, for every $s \in \{\log R, \ldots, R\}$, there is a deterministic online protocol that routes any permutation in time $O(\log_s N \cdot R)$ (i.e., with slowdown $O(\log_s N)$), if constant size buffers are available at each edge for storing packets, $O(s(d \log d + \log s) + \log N)$ space is available at each node, and $O(\log(s \cdot R))$ space is available at each packet for storing routing information.

Hence, for any bounded degree network of size N with routing number $R = \Omega(N^\epsilon)$, $\epsilon > 0$ constant (such as arbitrary planar networks of degree at most $N^{1-\epsilon'}$ for some constant $\epsilon' > 0$), $O(N^\delta)$ space suffices in the nodes for any constant $\delta > 0$ to obtain a routing time of $O(R)$ (i.e., a constant slowdown) for any permutation.

10. Introduction to Wormhole Routing

In the following three chapters we will concentrate on presenting wormhole routing strategies. Wormhole routing has several advantages over store-and-forward routing. In store-and-forward routing, if a b-flit message traverses a path of length d, and is never delayed, then it will reach its destination in bd steps (assuming that each channel can transmit one flit in each step). In a wormhole router, however, the first flit does not wait for the rest of the message. It therefore arrives at its destination after d steps, and the last flit of the message arrives after $d + b - 1$ steps. The difference in time is due to a better utilization of network edges by the wormhole router. In addition to reduced latency, wormhole routing also has the advantage that it can be implemented with small, fast switches, and is a realistic model for optical communication.

The primary drawback to wormhole routing is the contention that can occur even with moderate traffic, which leads to higher message latency. Whenever a message is unable to proceed due to contention, the header and data flits are not removed from the network. Instead, the message holds all channels it currently occupies. Since each of the channels along the path from the source to the destination is held from the time it is acquired until the entire message has traversed that channel, performance degradation due to contention can be severe and message latency can be unacceptably high. One suggestion (proposed by Dally [Da92] and others) was to allow a link to support several channels (in our terminology, to increase the link bandwidth), in order to reduce both latency and contention. We will show in the following that this strategy can in fact greatly improve the performance of wormhole routing strategies.

Before we do this, let us start with giving a historical overview of results in the area of wormhole routing, and presenting upper and lower bounds for wormhole routing in arbitrary path collections. In Chapter 11 we describe two universal oblivious wormhole routing protocols, and show how they can be applied to specific networks. Chapter 12 deals with all-optical routing. We present nearly tight bounds for the runtime of a very simple protocol for sending messages along an all-optical path collection. We further show how this protocol can be applied to specific networks.

10.1 History of Wormhole Routing

Wormhole routing has become the routing method of choice in the latest generation of massively parallel computers, appearing in experimental machines such as iWarp [BCC+88], the MIT J-machine [NWD93], and the Caltech MOSAIC, and commercial machines such as the Intel Paragon, Cray T3D [KFW94], Connection Machine CM-5 [LAD+92], and the nCUBE-2/3. It owes much of its recent popularity to an influential paper by Dally and Seitz [DS87], which introduced the method. Much of the paper by Dally and Seitz is devoted to the design of wormhole routing algorithms that avoid deadlocks. The solution to this problem in the paper by Dally and Seitz is to allow each link to emulate several virtual channels, and to construct a virtual network in which the worms can not form cycles. The virtual channels share the physical wire (or wires) provided by the link, but the switch maintains a separate buffer for each virtual channel. This solution has also been implemented in hardware. For example, each physical channel of the J-Machine supports 2 virtual channels [NWD93]. The paper by Dally and Seitz was followed by a large number of papers describing different forms of deadlock-free routing for various networks [PG91, CG94, Cy95, SJ95]. In the following we give an overview on results known about the runtime of wormhole routing in specific networks and universal oblivious wormhole routing protocols.

10.1.1 Routing in Specific Networks

Wormhole routing on butterfly networks has been studied extensively. Early work focused on the case that the link bandwidth is one. Kruskal and Snir [KS83] showed that if each input in an N-input butterfly sends a message to a randomly chosen output, then the expected number of worms that reach their destinations without ever being delayed is $\Theta(N/\log N)$. In Problem 3.285 of [Le92], Leighton describes an algorithm for solving a random routing problem in which each input has one worm of length L to send. The algorithm runs in $O((L + \log N)\log N)$ steps. In Problem 3.286, he shows that the algorithm can be converted to one that routes any permutation using Valiant's trick. The algorithm can easily be generalized to the case in which every input has k worms to send, and runs in $O(k(L+\log N)\log N)$ steps. For the interesting case of $L = O(\log N)$ and $k = \log N$, the time is $O(\log^3 N)$.

Felperin et al. [FRU92] independently discovered an algorithm for solving a random routing problem in which each input has one worm of length L to send. Their algorithm runs on a $\log N$-dimensional butterfly network in $O(L\log N \cdot \min\{L, \log N\})$ time, w.h.p., [FRU92]. Ranade, Schleimer and Wilkerson improved this result by showing that $q \cdot N$ worms of length $\log N$, q per input, can be routed to randomly chosen outputs in $O(q\log^2 N \cdot \log\log N)$ steps [RSW94]. Furthermore, they showed that $\Omega(q\log^2 N/(\log\log N)^2)$ steps are necessary for this task. Both algorithms used for proving the upper bounds above require the ability to buffer blocked worms [FRU92, RSW94].

Note that if we allow the flits of the worms to be routed independently, then several store-and-forward routing algorithms are known that can route q worms of length L in each input to randomly chosen outputs in in $O(q \cdot L + \log N)$ steps, w.h.p. (see [Le92]). Hence, the store-and-forward algorithms are faster. However, as the bandwidth of the links increases, the following results show that this difference gets smaller.

For $B > 1$, a non-linear dependence on B has been observed. The first result is due to Koch [Ko88], who showed that if each input in an N-input butterfly sends a worm to a randomly chosen output, then the expected number of worms that reach their destination without ever being delayed is $\Theta(N/(\log N)^{1/B})$ for any constant $B \geq 1$. Cypher et al. [CMSV96] found an algorithm for routing q worms form each input of an N-input butterfly to randomly chosen outputs in $O((q \cdot L \log^{1/B} N + (\log N + L) \log N)/B)$ steps, w.h.p. They also proved a lower bound of $\Omega(q \cdot L \log^{1/B} N \cdot (\log \log N)^{-2/B}/B)$ steps. Cole, Maggs, and Sitaraman [CMS96] independently discovered a randomized algorithm that routes any q-relation from the inputs to the outputs of an N-input butterfly in $O(L(q + \log N) \log^{1/B} N \cdot \log \log (qN)/B)$ steps. They also proved an $\Omega(L \cdot q(\min\{L, \log N\})^{1/B}/\log \log N)$ lower bound that holds for a broad class of algorithms.

Some results are also known for wormhole routing on meshes and tori. Felperin et al. [FRU92] analyze a simple algorithm for the $n \times n$ mesh. They show that it can route n^2 worms of length L, one per node, to random destinations in time $O(L \cdot n \log n + n^2/\log n)$, w.h.p., without buffering. This result was improved by Bar-Noy et al. [BRST93]. They present a randomized wormhole routing protocol for the $n \times n$ mesh that routes any permutation in $O(L \cdot n)$ steps, w.h.p., without buffering. This time bound is optimal since there exist permutations that need $\Omega(L \cdot n)$ steps. Recently, Bock [Bo96] showed, that for any d and L such that $L \cdot d^2 = O(\frac{n}{\log^2 n})$, there is an algorithm for routing n^d worms of length L in an (n, d)-torus, one per node, to random destinations in optimal time $O(L(d + n))$, w.h.p., using buffers that can store one flit.

10.1.2 Universal Routing

In the area of universal wormhole routing, Greenberg and Oh [GO93] created a randomized algorithm for arbitrary simple path collections of size n with link bandwidth 1, congestion C, and dilation D that requires time $O(C \cdot D \cdot \ell + C \cdot L \cdot \ell \log n)$, w.h.p., where $\ell = \min\{D, L\}$. This algorithm does require the ability to buffer blocked worms. Their result was improved by Cypher et al. [CMSV96]. In this paper a randomized algorithm is presented that routes worms of length L along an arbitrary simple path collection of size n with bandwidth B, congestion C, and dilation D in time $O((L \cdot C \cdot D^{1/B} + (D + L) \log n)/B)$, w.h.p., without buffering. Furthermore they present a randomized algorithm that routes worms of length L along an arbitrary leveled path collection of size n with bandwidth B, congestion C, and dilation D in time $O(L(D \log \log n + C + \log n \log \log n / \log \log(C \cdot D)))$, w.h.p., without buffering.

Before describing some of the protocols mentioned above, we summarize what has been found out so far about upper and lower bounds for wormhole routing in arbitrary networks.

10.2 Upper and Lower Bounds

Consider the problem of routing worms of length L along a path collection with congestion C, dilation D, and bandwidth B. The following result has been shown by Cole *et al.* [CMS96].

Theorem 10.2.1. *For every path collection with congestion C, dilation D, and bandwidth B there exists a protocol for routing worms of length L in*

$$\frac{(L+D)C(D \log D)^{1/B} 2^{O(\log^*(C/D))}}{B}$$

steps.

As for the offline protocols in the store-and-forward part, the proof uses as basic argument the Lovász Local Lemma. In [CMS96], the following lower bound can furthermore be found.

Theorem 10.2.2. *For any values of C, D, B, and L with $C, D \geq B + 1$ and $L = (1 + \Omega(1))D$, there is a path collection with congestion C, dilation D and bandwidth B such that routing worms of length L along these paths takes at least*

$$\Omega\left(\frac{L \cdot C \cdot D^{1/B}}{B}\right)$$

steps.

11. Oblivious Routing Protocols

In this chapter we present two universal oblivious wormhole routing protocols. In particular, we transform the trial-and-failure protocol and the duplication protocol presented in Section 7 into protocols that can be used for wormhole routing. We furthermore show how these protocols can be applied to specific networks. The results in this chapter have been taken from [CMSV96].

11.1 The Trial-and-Failure Protocol

In this section we describe a generalization of the trial-and-failure protocol to worms of arbitrary length L. We again use the following contention resolution rule.

> **B-priority rule:**
> If more than B worms attempt to use the same link during the same time step, then those B with lowest rank win.

The flits of those worms that lose are eliminated. Note that it can happen that a worm gets half through a link before flits of it get eliminated by a new worm. In this case the part of the worm that is not deleted is allowed to continue the routing. The protocol then works as follows.

Initially, each worm w_i chooses uniformly and independently from the other worms a random rank $r_i \in [K]$ (K will be specified later) and a random delay $d_i \in [D]$.

repeat

- **forward pass:** Each active worm w_i waits for d_i steps. Then it is routed along its path, obeying the B-priority rule.
- **backward pass:** For each worm that *completely* reached its destination during the forward pass, an acknowledgment is sent back to the source. Upon receipt of the acknowledgment, the source declares the worm inactive.

until no worm is active

Clearly, the forward pass needs $2D + L$ steps to be sure that every worm that has not been (partly) discarded during the routing completely reaches its destination. These worms are called *successful*. In the backward pass, the forward pass is run in reverse order. Therefore, no collisions can occur in the backward pass, and $2D$ steps suffice to send all acknowledgments back.

Theorem 11.1.1. *Suppose we are given an arbitrary simple path collection $\mathcal{P}$ of size n with link bandwidth B, congestion C, and dilation D. Further let $K \geq 4\ell C \cdot D^{-1+1/B}$, where $\ell = \min\{2L - 1, D\}$. Then the trial-and-failure protocol needs*

$$O\left(\frac{L \cdot C \cdot D^{1/B}}{B} + \left(\frac{\log n}{B} + 1\right)(D + L)\right)$$

steps, w.h.p., to route worms of length L along paths in $\mathcal{P}$ without using buffers.

Proof. The proof is analogous to the proof of Theorem 7.6.1 with the difference that the probability that two worms meet during the routing is now at most $\frac{\ell}{D}$ instead of $\frac{1}{D}$, where $\ell = \min\{2L - 1, D\}$.   □

This result is optimal for $B \geq \log D$ and $L \cdot C \geq (D + L)\log n$, or for $B \geq \log(nD)$. It has the following applications.

11.1.1 Wormhole Routing in Meshes and Tori

Applying Theorem 11.1.1 to the d-dimensional mesh with side length n (that is, with $N = n^d$ processors) yields $O(L \cdot n \cdot (n \cdot d)^{1/B}/B + (\frac{d\log n}{B} + 1)(n \cdot d + L))$ time, w.h.p., for routing n^d worms, one from each processor, to random destinations. This is because the expected congestion is at most n. A more careful analysis, tailored to meshes, gives the following improvement.

Theorem 11.1.2. *Given a d-dimensional mesh of side length n, there exists an on-line wormhole routing algorithm that routes $N = n^d$ worms of length L, one from each processor, to random destinations in*

$$O\left(\frac{L \cdot n \cdot d^{1/B}}{B} + \left(\frac{d\log n}{B} + 1\right)(n \cdot d + L)\right)$$

time, w.h.p., without buffering.

Proof. Let us choose the following routing strategy on a d-dimensional mesh with side length n. Given a random destination chosen independently and uniformly from the set of nodes, each worm chooses a random startup delay in $[d \cdot n]$ and afterwards is first routed according dimension 1, then according to dimension 2 and so on, until it reaches its destination.

In the following, A denotes an arbitrary linear array of length n in one dimension of the mesh. Let $E(A)$ be the expected number of worms that want to use links of A. Because of symmetry reasons, for uniformly and independently chosen destinations, $E(A)$ is the same for each linear array A. (Note that this argument is not true if we considered links instead of linear arrays.) Since the total number of linear arrays used by worms is at most $N \cdot d$ and the number of linear arrays is $d \cdot n^{d-1}$, $E(A)$ is at most $N \cdot d/(d \cdot n^{d-1}) = n$. Replacing the edge congestion by this array congestion in the proof of Theorem 11.1.1 yields the upper bound in Theorem 11.1.2.   □

Hence, for all constant-dimensional meshes with worms of length $\Omega(\log n)$, an asymptotically optimal runtime can be reached.

11.1.2 Wormhole Routing in Butterflies

The next theorem presents upper and lower bounds for wormhole routing on the d-dimensional butterfly network. We consider the problem of routing worms from the inputs on level 0 to the outputs on level d, where each of the $N = 2^d$ inputs has to send q worms to randomly selected outputs. Each worm has to follow the unique shortest path from its input to its output node (see the path system used for Theorem 7.2.4).

The upper bound is the same as that implied by Theorem 11.1.1. But whereas the algorithm used for proving Theorem 11.1.1 uses random ranks for priorities and assigns random delays (as, e.g., in [RSW94]), here we require neither random ranks nor delays. Instead, we give fixed, explicitly defined priorities to the worms.

The lower bound holds for any algorithm that moves the worms from the inputs to the outputs along their unique shortest paths without delaying them, even if offline routing is allowed. It extends the lower bound from [RSW94] to arbitrary bandwidth, and shows that our upper bound is nearly optimal.

Theorem 11.1.3. *Let $N = 2^d$. Given a d-dimensional butterfly network with link bandwidth $B \leq \log N$ and without buffers, routing qN worms, q from each input, to random outputs*

a) can be done by a deterministic online algorithm in

$$O\left(\frac{q \cdot L \cdot \log^{1/B} N + (\log N + L)\log N}{B}\right)$$

time, w.h.p., and
b) needs

$$\Omega\left(\frac{q \cdot L \cdot \log^{1/B} N}{B(\log\log N)^{2/B}}\right)$$

time, w.h.p., for $q \leq \log^\alpha N$ for any constant $\alpha > 0$.

Proof. Wir first prove part (a) of the Theorem.

Proof of Part (a): Suppose that qN worms, q per input, have to be sent to random outputs. Let $\ell = \min\{L, \log N\}$. Then the worms can be given delays such that the routing can be viewed as divided into $\lfloor \log N/\ell \rfloor$ independent passes in which every input has to send out at most $q' = \lceil q/\lfloor \log N/\ell \rfloor \rceil$ worms. Therfore, we have to show that there exists a rank allocation such that each pass takes $O(\frac{1}{B}(q' \cdot \log^{1/B} N + \log N))$ rounds, w.h.p., each requiring $2 \log N + L$ time steps. In the following we do not only show the existence but also describe a specific allocation that ensures the above routing time.

The ith worm ($0 \le i \le q'$) starting at node $(x_{\log N-1}, \ldots, x_0)$ on level 0 is assigned rank $i \cdot N + (x_0, \ldots, x_{\log N-1})_2$. Thus, the rank of a worm is essentially the bit-reversal number of its input node plus an offset. In the following we identify a worm with its rank. (Note that the ranks of all worms participating in the same phase are distinct.)

For an integer $m \ge 1$ with binary representation $(\ldots, m_2, m_1, m_0)$, we define

$$\delta(m) = \max\{i \le \log N \mid (m_{i-1}, \ldots, m_0) = 0^i\} .$$

$\delta(m)$ has the nice property that any two worms ω and ω' can not use the same edge before level $\log N - \delta(|\omega' - \omega|)$. Further the following two claims can be shown that will be used in the further analysis.

Claim 11.1.1. *For any $m, k \in \mathbb{N}$ with $k \le m$, there are at most $\binom{m}{k}/2^{i \cdot k}$ possibilities to choose a subset $\{x_1, \ldots, x_k\} \subseteq \{1, \ldots, m\}$ such that $\min\{\delta(x_1), \ldots, \delta(x_k)\} = i$.*

Proof. Let $m \in \mathbb{N}$ be an arbitrary constant. According to the definition of $\delta(i)$, for every $i \in \{1, \ldots, \lfloor \log m \rfloor\}$ there are at most $\lfloor m/2^i \rfloor$ numbers $x \in \{1, \ldots, m\}$ with $\delta(x) \ge i$. Therefore there are at most $\binom{\lfloor m/2^i \rfloor}{k}$ choices for $\{x_1, \ldots, x_k\}$. Since for all $j \in \{0, \ldots, k-1\}$ it holds that $\frac{m/2^i - j}{m - j} \le \frac{1}{2^i}$ we get $\binom{\lfloor m/2^i \rfloor}{k} \le \frac{1}{2^i}\binom{m}{k}$. $\qquad \square$

Claim 11.1.2. *For any $k \in \mathbb{N}$ and any strictly decreasing function $f : \mathbb{N} \to \mathbb{R}$ it holds that*

$$\sum_{i=1}^{k} 2^{\delta(i)} f(i) \le (\log N + 1) \sum_{i=1}^{k} f(i) .$$

Proof. Let $k \in \mathbb{N}$ be an arbitrary constant. We first show that $\sum_{i=1}^{k'} 2^{\delta(i)} \le (\log N + 1)k'$ for any $k' \in \{1, \ldots, k\}$. Since there are at most $\lfloor k'/2^i \rfloor$ numbers $x \in \{1, \ldots, k'\}$ with $\delta(x) = i$, it holds

$$\sum_{i=1}^{k'} 2^{\delta(i)} \le \sum_{i=1}^{\lfloor \log k' \rfloor} \left\lfloor \frac{k'}{2^i} \right\rfloor 2^{\min\{i, \log N\}}$$

$$\leq \quad \sum_{i=1}^{\log N} k' + \sum_{i=\log N+1}^{\lfloor \log k' \rfloor} \left\lfloor \frac{k'}{2^i} \right\rfloor \cdot N$$

$$\leq \quad (\log N + 1)k' \tag{11.1}$$

Since f is strictly decreasing, we can show by induction that

$$\sum_{i=1}^{k'} 2^{\delta(i)} f(i) \leq (\log N + 1) \sum_{i=1}^{k'} f(i)$$

for all $k' \geq 1$. The case $k' = 1$ is trivial. Suppose that $k' \geq 1$ and $\delta(k'+1) = m$. If $m \leq \log N$ then

$$\sum_{i=1}^{k'+1} 2^{\delta(i)} f(i) \quad \overset{(11.1)}{\leq} \quad (\log N + 1) \sum_{i=1}^{k'} f(i) + (\log N + 1)f(k' + 1)$$

$$= \quad (\log N + 1) \sum_{i=1}^{k'+1} f(i) \; .$$

Suppose now that $m > \log N$. Then we get from Inequality (11.1) that $\sum_{i=1}^{k'} 2^{\delta(i)} \leq (\log N + 1)k' - (m - (\log N + 1))$. Hence it holds

$$\sum_{i=1}^{k'+1} 2^{\delta(i)} f(i) \quad \leq \quad (\log N + 1) \sum_{i=1}^{k'} 2^{\delta(i)} f(i) +$$

$$(\log N + 1 - m)f(k') + m \cdot f(k' + 1)$$

$$\leq \quad (\log N + 1) \sum_{i=1}^{k'+1} 2^{\delta(i)} f(i)$$

From this the claim follows. □

Suppose $\omega_1 < \ldots < \omega_{B+1}$ are $B + 1$ worms. Then there are

$$N \cdot \left(\lfloor 2^{\min\{\delta(\omega_2 - \omega_1),\ldots,\delta(\omega_{B+1}-\omega_1)\}-1} \rfloor \right)^B$$

ways to choose the destinations of the worms such that all $B + 1$ routing paths share an edge, i.e. the worms can not reach their destination altogether in the same round. This is because the worms can not meet before level $\ell = \log N - \min\{\delta(\omega_2 - \omega_1), \ldots, \delta(\omega_{B+1} - \omega_1)\}$, and once the destination of worm ω_1 is fixed, the node in that level and the edge leaving it are fixed, which means, that only $\lfloor 2^{\ell-1} \rfloor$ destinations can be chosen by the B other worms.

Now, consider an arbitrary worm ω_1 with some fixed destination, and let m be an arbitrary integer. Then the number of possibilities to choose B

worms $\omega_2, \ldots, \omega_{B+1}$ and their destinations such that $\omega_1 < \omega_2 < \ldots < \omega_B < \omega_{B+1} = \omega_1 + m$, and all $B + 1$ routing paths share an edge, is at most

$$\sum_{\substack{\omega_2,\ldots,\omega_B \\ \omega_1 < \omega_2 < \ldots < \omega_1 + m}} 2^{(\min\{\delta(\omega_2-\omega_1),\ldots,\delta(\omega_B-\omega_1),\delta(m)\}-1)\cdot B}$$

$$\leq \binom{m}{B-1} \left[\sum_{i=0}^{\delta(m)-1} 2^{-i(B-1)} \cdot 2^{(i-1)B} + \sum_{i=\delta(m)}^{\infty} 2^{-i(B-1)} \cdot 2^{(\delta(m)-1)\cdot B} \right]$$

$$\leq \binom{m}{B-1} 2^{\delta(m)} . \tag{11.2}$$

This is because according to Claim 11.1.1 for every level $\log N - i$ there are at most $\binom{m}{B-1}/2^{i(B-1)}$ possibilities to choose $B - 1$ worms $\omega_2, \ldots, \omega_B$ that share an edge leaving that level, and $2^{\min\{i,\delta(m)\}-1}$ possibilities to choose a destination for each of these worms and ω_{B+1}.

Now we are able to estimate the number of (s, B)-delay sequences. For this purpose we have to count the number of ways to choose $s \cdot B + 1$ worms $\omega_1, \ldots, \omega_{s \cdot B+1}$ and their destinations such that $0 \leq \omega_1 < \omega_2 < \ldots < \omega_{s \cdot B+1} < q'N$ and such that for every $1 \leq i \leq s$, the routing paths of the worms $\omega_{i \cdot B+1}, \omega_{i \cdot B+2}, \ldots, \omega_{(i+1) \cdot B+1}$ share an edge. Of course, there are at most $q'N^2$ possibilities to choose ω_1 and its destination. Define $R = q'N$, and let $m_i = \omega_{i \cdot B+1} - \omega_{(i-1) \cdot B+1}$. Then it follows by repeatedly applying Inequality (11.2) that the number of (s, B)-delay sequences is at most

$$q'N^2 \cdot \sum_{\substack{m_1,m_2,\ldots,m_s \geq 1 \\ \sum m_i \leq R}} \prod_{i=1}^{s} \binom{m_i}{B-1} \cdot 2^{\delta(m_i)}$$

$$\leq q'N^2 \cdot \binom{R}{s(B-1)} \underbrace{\sum_{\substack{m_1,\ldots,m_s \geq 1 \\ \sum m_i \leq R}} \prod_{i=1}^{s} 2^{\delta(m_i)}}_{=:A_{s,R}}$$

We now show by induction on s that $A_{s,R} \leq (\log N + 1)^s \binom{R}{s}$. For $s = 0$, this inequality trivially holds. For $s \geq 1$, we have

$$A_{s,R} = \sum_{m_s=1}^{R-1} 2^{\delta(m_s)} \sum_{\substack{m_1,m_2,\ldots,m_{s-1} \geq 1 \\ \sum m_i \leq R-m_s-1}} \prod_{i=1}^{s-1} 2^{\delta(m_i)}$$

$$= \sum_{m=1}^{R-1} 2^{\delta(m)} \cdot A_{s-1,R-m}$$

$$\leq \sum_{m=1}^{R-1} 2^{\delta(m)} \cdot (\log N + 1)^{s-1} \binom{R-m}{s-1}$$

$$\overset{\text{Claim 11.1.2}}{\leq} \quad (\log N + 1)^s \sum_{m=1}^{R-1} \binom{R-m}{s-1}$$

$$= \quad (\log N + 1)^s \binom{R}{s} .$$

Since each worm chooses uniformly and independently at random one out of N possible destinations, the probability that a worm is still active after

$$T \geq \frac{4e \cdot q' \cdot (\log N + 1)^{1/B} + (\alpha + 2) \log(q'N)}{B}$$

$$= O\left(\frac{q' \cdot (\log N + 1)^{1/B} + \log N}{B}\right)$$

rounds can be bounded by

$$q'N^2 \cdot \left(\frac{R}{T(B-1)}\right) \cdot (\log N + 1)^T \cdot \binom{R}{T} \cdot N^{-TB}$$

$$\leq \quad q'N^2 \cdot \left(\frac{2e \cdot q'N \cdot (\log N + 1)^{1/B}}{TB \cdot N}\right)^{TB}$$

$$\leq \quad (q' \cdot N)^{-\alpha} .$$

This yields part (a) of Theorem 11.1.3.

Proof of part (b): Consider the problem of routing $\log^\alpha N$ worms from each input node to random output nodes for any constant $\alpha > 0$. Let us choose an arbitrary subset of these worms of size $m \geq 7BN/\log N$. We first bound the probability P that all these worms can be routed simultaneously, i.e., they can arrive at their random destinations in $\log N + L$ steps. In the following we say that worms *collide* if more than B of them want to travel along the same edge at the same time.

In order to cope with dependencies between the collision probabilities on different levels we use an idea from Ranade *et al.* [RSW94] which is to divide the butterfly into many small subbutterflies. Let $k = \lceil \log \log N \rceil$ and $K = 2^k$. Then the first k levels consist of N/K small distinct subbutterflies. We denote the number of worms starting in the ith of these subbutterflies by m_i.

We first estimate the probability P_i that the m_i worms do not collide in one of the $2K$ edges leaving subbutterfly i. Let $e_1, \ldots, e_{2K}$ denote these edges. Further, let E_j denote the event that more than B worms want to travel along the edge e_j, and let E'_j denote the event that exactly $B + 1$ worms want to travel along the edge e_j. If $B + 1 \leq m_i \leq 2KB$ then we get

$$\Pr[E_j] \quad \geq \quad \Pr[E'_j]$$

$$= \quad \binom{m_i}{B+1}\left(\frac{1}{2K}\right)^{B+1}\left(1 - \frac{1}{2K}\right)^{m_i - B - 1}$$

$$\overset{B+1 \leq m_i}{\geq} \left(\frac{m_i}{2K(B+1)}\right)^{B+1} 3^{-(m_i-B-1)/2K}$$

$$\overset{m_i \leq 2KB}{\geq} \left(\frac{m_i}{2K(B+1)}\right)^{B+1} \cdot 3^{-B} \quad,$$

and therefore,

$$P_i \leq \Pr\left[\bigwedge_{j=1}^{2K} \neg E_j\right]$$

$$\leq \prod_{j=1}^{2K} \Pr\left[\neg E_j \middle| \bigwedge_{\ell=1}^{j-1} \neg E_\ell\right] \leq \prod_{j=1}^{2K} \Pr[\neg E_j]$$

$$\leq \left(1 - \left(\frac{m_i}{2K(B+1)}\right)^{B+1} \cdot 3^{-B}\right)^{2K}$$

$$\leq \exp\left(-m_i^{B+1} \cdot (B+1)^{-(B+1)} \cdot (6K)^{-B}\right) \quad.$$

Now we can bound the probability P that in every subbutterfly i the m_i worms do not collide at edges leaving that subbutterfly. If there is at least one subbutterfly in which more than $2KB$ worms start, then we have $P = 0$. Otherwise,

$$P \leq \prod_{\substack{i=1 \\ m_i \geq B+1}}^{N/K} P_i$$

$$\leq \exp\left(-\left(\sum_{\substack{i=1 \\ m_i \geq B+1}}^{N/K} m_i^{B+1}\right) \cdot (B+1)^{-(B+1)} \cdot (6K)^{-B}\right).$$

Let ℓ denote the number of subbutterflies in which more than B worms start, and let m' denote the number of worms that start in subbutterflies with more than B worms. Then

$$m' = \sum_{\substack{i=1 \\ m_i \geq B+1}}^{N/K} m_i \geq m - \frac{BN}{K} \geq \frac{6}{7}m \quad.$$

Hence we get

$$\sum_{\substack{i=1 \\ m_i \geq B+1}}^{N/K} m_i^{B+1} \geq \sum_{\substack{i=1 \\ m_i \geq B+1}}^{N/K} \left(\frac{m'}{\ell}\right)^{B+1} \geq \ell \left(\frac{6m}{7\ell}\right)^{B+1} \quad.$$

Thus, we can conclude that the probability that collisions occur at edges leaving level k of the butterfly is at most

$$P \quad \leq \quad \exp\left(-\ell\left(\frac{6m}{7\ell}\right)^{B+1} \cdot (B+1)^{-(B+1)} \cdot (6K)^{-B}\right)$$

$$\overset{\ell \leq N/K}{\leq} \quad \exp\left(-(6/7) \cdot (7N)^{-B} \cdot m^{B+1} \cdot (B+1)^{-(B+1)}\right) \quad .$$

We can repeat this argument to obtain upper bounds for the probability of having no collisions in edges leaving levels $2k$, $3k$, $4k$ and so on. Since we obtain all these bounds by using only information regarding the k levels in front of the respective collision level, the probabilities for each of these levels can be considered as independent. Hence the probability that the m worms do not collide on their way through the whole butterfly is at most

$$\left(\exp\left(-(6/7) \cdot (7N)^{-B} \cdot m^{B+1} \cdot (B+1)^{-(B+1)}\right)\right)^{\left\lfloor \frac{\log N}{k} \right\rfloor}$$

$$\leq \quad \exp\left(-\frac{\log N \cdot m^{B+1}}{2\log\log N \cdot (7N)^B \cdot (B+1)^{B+1}}\right) .$$

Therefore, the probability that there exists a set of

$$m \geq 7N \cdot \left(\frac{2(\alpha+2)\log^2 \log N}{\log N} \cdot (B+1)^{B+1}\right)^{1/B}$$

worms that can be routed simultaneously is at most

$$\binom{N \log^\alpha N}{m} \cdot \exp\left(-\frac{\log N \cdot m^{B+1}}{2\log\log N \cdot (7N)^B \cdot (B+1)^{B+1}}\right)$$

$$\leq \quad \exp\left(m\left((\alpha+1)\log\log N - \frac{\log N \cdot m^B}{2\log\log N \cdot (7N)^B \cdot (B+1)^{B+1}}\right)\right)$$

$$\leq \quad \exp(-m) .$$

As a consequence, the routing takes at least time

$$\frac{LN \log^\alpha N}{m} = \Omega\left(\frac{L(\log N)^\alpha}{B}\left(\frac{\log N}{\log^2 \log N}\right)^{1/B}\right)$$

with probability at least $1 - \exp(-m)$. This proves part (b) of Theorem 11.1.3.
□

11.1.3 Further Applications

Applying Theorem 11.1.1 to node- and edge-symmetric networks gives the following results.

Corollary 11.1.1. *Given any node-symmetric network of size N, diameter D, and link bandwidth B there exists an online wormhole routing algorithm* *that routes N worms of length L, one from each processor, to random destinations in*

$$O\left(\frac{L \cdot D^{1+1/B}}{B} + \left(\frac{\log N}{B} + 1\right)(D + L)\right)$$

time, w.h.p., without buffering.

Corollary 11.1.2. *Given any edge-symmetric network of size N, diameter D, degree d, and link bandwidth B there exists an online wormhole routing algorithm that routes n worms of length L, one from each processor, to random destinations in*

$$O\left(\frac{L \cdot D^{1+1/B}}{d \cdot B} + \left(\frac{\log N}{B} + 1\right)(D + L)\right)$$

time, w.h.p., without buffering.

Applying Corollary 5.3.2 and Theorem 5.1.1 yields the following result.

Corollary 11.1.3. *Given any bounded degree expander of size N, there exists an online wormhole routing algorithm that routes N worms of length L, one from each processor, to random destinations in*

$$O\left(\frac{L \log^{1+1/B} N}{B} + \left(\frac{\log N}{B} + 1\right)(\log N + L)\right)$$

time, w.h.p., without buffering.

11.2 The Duplication Protocol

The duplication protocol presented in Section 7.7 can easily be adapted to the case of routing worms in arbitrary leveled networks. To be able to use the same analysis, we require the worms to collide with each other only by their heads. That is, if we want to route worms of length L, then their delays have to be a multiple of L. Then we get the following result.

Theorem 11.2.1. *Given any leveled path collection $\mathcal{P}$ of size n with congestion C, dilation D, and bandwidth $\Theta(\log(C{\cdot}D)/\log\log(C{\cdot}D))$, the duplication protocol requires*

$$O\left(L\left(D\log\log n + C + \frac{\log n \cdot \log\log n}{\log\log(C \cdot D)}\right)\right)$$

time, w.h.p., to route worms of length L along paths in $\mathcal{P}$ without buffering.

For example, if $D \geq \log n$ and $C \geq D\log\log n$, Theorem 11.2.1 yields a routing time of $O(L \cdot C)$, which is only a factor of $\log C/\log\log C$ away from the lower bound. For d-dimensional meshes and tori, combining Theorem 11.2.1 with techniques in [LMRR94] yields the following result.

Corollary 11.2.1. *Given any d-dimensional mesh of side length n and link bandwidth $B = \Theta(\log(dn)/\log\log(dn))$ there exists an online wormhole routing algorithm that routes $N = n^d$ worms of length L, one from each processor, to random destinations in*

$$O\left(L\left(d \cdot n \log\log N + \frac{\log N \log\log N}{\log\log\log(d \cdot n)}\right)\right)$$

time, w.h.p., without buffering.

For butterflies, Theorem 11.2.1 yields the following result.

Corollary 11.2.2. *Given a butterfly with N inputs and link bandwidth $\Theta(\frac{\log\log N}{\log\log\log N})$, routing qN worms, q from each input, to random outputs can be done in time*

$$O\left(L(q + \log N \cdot \log\log N)\right) \;,$$

w.h.p., without buffering.

11.3 Summary of Main Results

At the end of this chapter, let us summarize its main results. After presenting upper and lower bounds for offline wormhole routing, we presented two online protocols: the trial-and-failure protocol and the duplication protocol. In particular, we could prove the following two results (see Theorem 11.1.1 and Theorem 11.2.1).

Suppose we are given an arbitrary simple path collection $\mathcal{P}$ of size n with link bandwidth B, congestion C, and dilation D. Further let $K \geq 4e\ell C \cdot D^{-1+1/B}$, where $\ell = \min\{2L - 1, D\}$. Then the trial-and-failure protocol needs

$$O\left(\frac{L \cdot C \cdot D^{1/B}}{B} + \left(\frac{\log n}{B} + 1\right)(D + L)\right)$$

steps, w.h.p., to route worms of length L along paths in $\mathcal{P}$ without using buffers.

Given any leveled path collection $\mathcal{P}$ of size n with congestion C, dilation D, and bandwidth $\Theta(\log(C \cdot D)/\log\log(C \cdot D))$, the duplication protocol requires

$$O\left(L\left(D \log\log n + C + \frac{\log n \cdot \log\log n}{\log\log(C \cdot D)}\right)\right)$$

time, w.h.p., to route worms of length L along paths in $\mathcal{P}$ without buffering.

We further showed how to improve the time bound of the trial-and-failure protocol when applied to d-dimensional meshes, and how to remove random ranks and delays when applied to butterflies.

12. Protocols for All-Optical Networks

In this chapter we present analyses of variants of the trial-and-failure protocol that can be run efficiently on an emerging generation of networks known as *all-optical networks* (see, e.g., [Br90, CNW90, Gr92, Pe83, Ra93, IEEE94, Ch95]). These networks promise data transmission rates several orders of magnitudes higher than current networks. The key to high speeds in these networks is to maintain the signal in optical form, thereby avoiding the prohibitive overhead of conversion to and from the electrical form. (Traditional networks use the electrical form to switch signals along routes, and to restore signals. Signals can be modulated electronically at a maximum bit rate of the order of tens of Gps, while the optical fiber bandwidth is at least 25 THz [Ch95]. The high bandwidth of the optical fiber is utilized through WDM (wavelength-division multiplexing, see Section 2.1): two signals connecting different source-destination pairs may share a link, provided they use different wavelengths of light. The major applications for such networks are in video conferencing, scientific visualization and real-time medical imaging, high-speed supercomputing and distributed computing [Gr92, Ra93, DV93].

12.1 An All-Optical Hardware Model

In the following we consider routing elements that are capable of directing messages at different wavelengths to different destinations and detecting collisions of messages. A routing element (or *router* in short) consists of wavelength-selective *switches* and *couplers*.

The task of the switches is to direct different wavelengths to different directions. Several types of optical switches have already been developed [HC+93, BCFR96].

The task of the couplers is to combine the signals from many incoming optical fibers into one outgoing optical fiber. Since we do not allow central control, collisions might occur, that is, two or more signals from different incoming fibers use the same wavelength. In our design of protocols we will consider two different strategies to avoid collisions:

– If a message that arrives at a coupler uses a wavelength already used by another message traversing the coupler, the new message is eliminated.

This can be realized with the help of detector arrays that tell the electronic control of the coupler which wavelengths are currently used, and wavelength-selective filters at each incoming fiber.
- If a message that arrives at a coupler uses a wavelength already used by another message traversing the coupler, the message with higher priority is forwarded and the other suspended. This strategy is significantly more difficult to realize than the first strategy. We consider it nevertheless to see whether it would be worth to invest this effort or better use the first type.

We call a coupler using the first rule *serve-first coupler* and *priority coupler* otherwise.

The following picture illustrates how a 2×2 router can be built by switches and couplers.

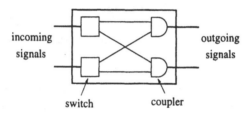

Fig. 12.1. A 2×2 router.

The number of wavelengths a router can handle is called the *bandwidth* of the router and denoted by B. As defined for the coupler above, we distinguish between two rules to avoid collisions in the router: the *serve-first rule* and the *priority rule*.

12.2 Overview on All-Optical Routing

All-optical routing problems have been considered for two basic network models: the *non-reconfigurable* or *switchless* networks, and the *reconfigurable* networks. In the first class of networks, a fixed set of wavelengths is assigned to every connection between any input and output of a router, whereas in the second class switches are allowed, that is, connections between the inputs and the outputs of a router can change the set of wavelengths that are supported by them. The *elementary* switch can not direct different wavelengths arriving at some input to different outputs, whereas the *generalized* switch can do this. Figure 12.2 gives an example of a router for non-reconfigurable networks and reconfigurable networks with elementary switches.

The routing protocols developed for the class of non-reconfigurable networks can be separated into two categories: the *single hop* strategies and

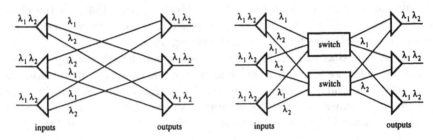

Fig. 12.2. Structure of a switchless (left) and an elementary (right) router.

the *multi hop* strategies. In a single hop strategy, messages are not allowed to change their wavelength somewhere along their routing path, while in a multi hop strategy they are usually allowed to do so for a bounded number of times (each time is referred to as a *hop*).

For the class of single hop strategies, Barry and Humblet [BH92] could prove that, for any network, permutation routing requires $\Omega(\sqrt{n})$ wavelengths, where n is the number of nodes in the network. They also showed that oblivious permutation routing can be done using $\lceil n/2 \rceil + 2$ wavelengths. Awerbuch *et al.* [ABC94] could prove the existence of a switchless permutation network using $O(\sqrt{n \log n})$ wavelengths. They also show how to construct a switchless permutation network using $O(\sqrt{n} 2^{(\log n)^{0.8+o(1)}})$ wavelengths.

Multi hop strategies have been considered, e.g., by [CGK92, CGK93, ZA95, GZ94a, GZ94b, KKP97]. Chlamtac *et al.* [CGK92] study the problem of establishing multi hop paths for a given static and dynamic set of circuit demands. In another paper, Chlamtac *et al.* [CGK93] considered the problem of embedding regular networks into the original fiber topology. They present bounds on the number of wavelengths required to simulate a regular topology. Furthermore, algorithms for embedding different regular topologies are described and their performances are evaluated. Zhang and Acampora [ZA95] follow this line by studying two heuristic algorithms for embedding the hypercube into a physical fiber topology. Gerstel and Zaks [GZ94a, GZ94b] study layouts for chains, rings, meshes and trees. Kranakis *et al.* [KKP97] give asymptotically tight bounds on the number of hops required for the chain and the mesh given the number of available wavelengths (there expressed as congestion).

Within the field of reconfigurable networks, a number of papers [BS91, CNW90, MS93] have formulated the routing problem for both elementary and generalized switches as combinatorial optimization problems. For networks with w available wavelengths and elemtary switches, Barry and Humblet [BH94] showed that the number of 2×2 switches required to support permutation routing is $\Omega(n \log(n/w^2))$. Awerbuch *et al.* [ABC94] could show the existence of a permutation network using $O(n \log \frac{n \log w}{w^2})$ switches and constructed a permutation network using $O(n \log(2^{(\log w)^{0.8+o(1)}}/w^2))$ switches.

When the transmitters are fixed-tuned and the receivers are tunable, Pieris and Sasaki [PS93] showed that the number of 2×2 switches required for permutation routing is $\Omega(n \log(n/w))$, and constructed such a network using $O(n \log(n/w))$ switches.

When generalized switches are used, Pankaj et al. [Pa92, PG95] showed that $\Omega(\log n)$ wavelengths are required for permutation routing. They also showed that rearrangeably non-blocking permutation routing can be done with $O(\log^2 n)$ wavelengths and wide-sense non-blocking permutation routing can be done with $O(\log^3 n)$ wavelengths in popular interconnection networks such as the shuffle exchange network, the DeBruijn network, and the hypercube. Awerbuch et al. could show a tight bound of $\Theta(\log n)$ wavelengths for both rearrangeable and wide-sense non-blocking permutation networks.

Raghavan and Upfal [RU94] prove results that establish a connection between the expansion of a network and the number of wavelengths required for routing on it, considering both elementary and generalized switches. In [RS95], Ramaswami and Sivarajan present a lower bound on the blocking probability for any so-called routing and wavelength assignment (RWA) algorithm if requests and terminations of connections arrive at random, and generalized switches are used. They study both the case that wavelength conversion is allowed and not allowed at the routers.

In case that wavelength conversion is allowed at every router, the trial-and-failure protocol and duplication protocol presented in the previous chapter can be used to obtain a fast routing time, w.h.p. However, all-optical devices for wavelength conversion are still a topic in research and might significantly increase the cost of a router. Therefore we will show in this chapter how far one can get without wavelength conversion.

12.3 A Simple, Efficient Protocol

In the following we investigate how much time is necessary to route messages to their destinations given an arbitrary shortcut-free path collection with links of some fixed bandwidth in case that wavelength conversion is not allowed. For this we need the following definition.

The *path congestion* $\tilde{C}$ of a path collection $\mathcal{P}$ is defined as the maximum over all paths p in $\mathcal{P}$ of the total number of paths that share an edge with p.

Since the communication time is usually much higher than the calculation time of processors (this is the case even in all-optical networks, mainly caused by the time to initiate a communication) it is very important to have routing protocols that are as simple as possible. Hence the processors should use strategies that do not rely on any coordination. Since messages can not be buffered during the routing along their path there are basically two types of strategies to handle such a situation: starting messages with random delays, or assigning priorities to messages. Clearly, the most simple protocol that can be thought of for sending worms along a fixed path collection using routers with bandwidth B is the following variant of the trial-and-failure protocol presented earlier.

Trial-and-Failure Protocol:
all n worms are declared active
for $t = 1$ to T do:

- each active worm is sent out from its source with random startup delay in some suitably chosen range $[\Delta_t]$ using a random wavelength in $[B]$
- for every worm that completely reaches its destination, an acknowledgement is sent back to the source immediately afterwards
- every source that gets back an acknowledgement declares its worm as inactive

Let us call the execution of one for-loop one *round*. Clearly, round t requires at most $\Delta_t + 2(D + L)$ steps to be sure that either an acknowledgement of a successful worm reaches its source, or the worm or its acknowledgement has been (partly) discarded. (Note that if we use priority routers it can happen that worms are only partly discarded.)

Previously, only delay sequence arguments were used to analyze such protocols (see, e.g., [CMSV96, SV96]). In this chapter we use delay tree arguments that yield much more accurate upper bounds on the runtime. In particular, we are able to prove the following three results depending on the contention resolution rule. Their proofs can be found in Section 12.4 and Section 12.5 and have been taken from [FS97]. Let $\alpha = \tilde{C} + B(\frac{D}{L} + 1) + 2$ and $\beta = \alpha/\tilde{C} + 2$. The first theorem presents a nearly tight analysis of the protocol above for leveled path collections.

Theorem 12.3.1. *For any leveled path collection of size n with dilation D and path congestion $\tilde{C}$ using serve-first routers with bandwidth B the protocol above routes a worm of length L along each of these paths in time*

$$O\left(\frac{L \cdot \tilde{C}}{B} + \left(\sqrt{\log_\alpha n} + \log\log_\beta n \right) \left(\frac{L \log n}{B} + D + L \right) \right),$$

w.h.p. Furthermore there exists a leveled path collection such that, for any $L \geq 2$, the expected runtime is bounded by

$$\Omega\left(\frac{L \cdot \tilde{C}}{B} + \left(\sqrt{\log_\alpha n} + \log\log_\beta n \right) (D + L) \right).$$

Since in contrast to leveled path collections it can happen in some shortcut-free path collections that worms prevent each other from reaching their destinations, we get a slightly worse result for arbitrary shortcut-free path collections.

Theorem 12.3.2. *For any shortcut-free path collection of size n with dilation D and path congestion $\tilde{C}$ using serve-first routers with bandwidth B the protocol above routes a worm of length L along each of these paths in time*

$$O\left(\frac{L \cdot \tilde{C}}{B} + (\log_\alpha n + \log\log_\beta n)\left(\frac{L\log^{3/2} n}{B} + D + L\right)\right),$$

w.h.p. Furthermore there exists a shortcut-free path collection such that, for any $L \geq 2$, the expected runtime is bounded by

$$\Omega\left(\frac{L \cdot \tilde{C}}{B} + (\log_\alpha n + \log\log_\beta n)(D + L)\right).$$

As will become clear in the proof of Main Theorem 12.3.2, for the case $L = 1$ or there are no directed loops in the path collection of length below $\sqrt{\log_\alpha n}$, the upper bound in Main Theorem 12.3.2 can be reduced to the upper bound in Main Theorem 12.3.1. For any other situation, we also obtain this bound if we replace the serve-first routers by priority routers.

Theorem 12.3.3. *For any collection of n shortcut-free paths with dilation D and path congestion $\tilde{C}$ using priority routers with bandwidth B the protocol above routes a worm of length L along each of these paths in time*

$$O\left(\frac{L \cdot \tilde{C}}{B} + \left(\sqrt{\log_\alpha n} + \log\log_\beta n\right)\left(\frac{L\log n}{B} + D + L\right)\right),$$

w.h.p. Furthermore there is a shortcut-free path collection and a strategy for assigning priorities to the worms such that, for any $L \geq 2$, the expected runtime is bounded by

$$\Omega\left(\frac{L \cdot \tilde{C}}{B} + \left(\sqrt{\log_\alpha n} + \log\log_\beta n\right)(D + L)\right).$$

Note that the upper bound holds for *any* assignment of priorities to the worms such that no two worms with the same priority can meet in one round, whether these priorities are changed from round to round, chosen randomly, or deterministically.

The theorems indicate that for shortcut-free path collections the priority rule is more powerful than the serve-first rule. Often, $\Omega(\frac{L \cdot \tilde{C}}{B} + D + L)$ is a lower bound for any protocol using serve-first or priority routers. In this case the runtime of our protocol can get optimal if $\tilde{C}$ is large enough compared to D and L. Note that, for instance, for the butterfly network of size N the average path congestion of permutation routing problems is $\Theta(\log^2 N)$, whereas its diameter is $O(\log N)$.

The upper and lower bounds in Theorem 12.3.1 and 12.3.3 will be proved in Section 12.4, and the upper and lower bound in Theorem 12.3.2 will be given in Section 12.5. In the following, we describe some applications of the trial-and-failure protocol.

12.3.1 Applications

The results presented above can be applied, e.g., to node-symmetric networks.

Theorem 12.3.4. *For any bounded degree node-symmetric network of size n with diameter D using priority routers with bandwidth B there is an online protocol for routing a randomly chosen function in time*

$$O\left(\frac{L \cdot D^2}{B} + \left(\sqrt{\log_D n} + \log\log n\right)(D + L)\right),$$

w.h.p.

Proof. In Theorem 5.1.2 it is shown that for every node-symmetric network with diameter D there exists a shortcut-free path system with dilation at most D and expected congestion at most D. Using Chernoff bounds we therefore get that a randomly chosen function has a path collection with path congestion $O(D^2 + \log n)$, w.h.p., where n is the size of the network. Using this in the time bound of Theorem 12.3.3 yields the theorem. $\qquad\square$

Note that for $B = 1$ this time bound is much better than the time bound in Theorem 11.1.1. (For $B > 1$ we can not compare the results since Theorem 11.1.1 allows wavelength conversion which we do not allow here.) The result in Theorem 12.3.4 can be improved for d-dimensional meshes and tori.

Theorem 12.3.5. *For any d-dimensional mesh of side length n using serve-first routers with bandwidth B there is an online protocol for routing a randomly chosen function in time*

$$O\left(\frac{L \cdot d \cdot n}{B} + (\sqrt{d} + \log\log n)\left(\frac{L \cdot d\log n}{B} + d \cdot n + L\right)\right),$$

w.h.p.

Proof. From Theorem 11.1.2 it follows that there exists a routing strategy for routing a randomly chosen function that has a path congestion of $O(d \cdot n)$, w.h.p., and in which cyclic eliminations of worms can not appear. Since the size N of a d-dimensional mesh with side length n is equal to n^d, it follows that

$$\sqrt{\log_\alpha N} = O\left(\sqrt{d\log_{dn} n}\right) = O\left(\sqrt{d}\right),$$

where α is chosen as in the theorems above. In case that $\sqrt{d} \leq \log\log N$ we have that $n \geq N^{1/\log\log N}$ and therefore $\log\log N = O(\log\log n)$. This concludes the proof. $\qquad\square$

This improves the time bound in Theorem 11.1.2 for the case that $B = 1$. Note that Theorem 11.1.2 requires the worms to have ranks, which we do not require here any more. In case that we use butterfly networks, we obtain the following result.

Theorem 12.3.6. *For any $\log n$-dimensional butterfly using serve-first routers with bandwidth B there is a leveled path system such that a randomly chosen q-function can be routed from the inputs to the outputs in time*

$$O\left(\frac{L \cdot q \log n}{B} + \sqrt{\frac{\log n}{\log(q \log n)}}\left(\frac{L \log n}{B} + L + \log n\right)\right),$$

w.h.p.

For $B = 1$, this improves for some cases the time bound in Theorem 11.1.3.

12.4 Proof of Theorems 12.3.1 and 12.3.3

In this section we prove upper and lower bounds on the runtime of our protocol using serve-first routers in leveled path collections, or priority routers in shortcut-free path collections. In order to simplify the presentation, we will concentrate on serve-first routers in leveled path collections, and note the analogy to routing with priority routers in shortcut-free path collections whenever it is necessary.

Hence suppose we want to route worms of length L along a collection of n leveled paths with path congestion $\tilde{C}$ and dilation D, using serve-first routers with bandwidth B. (In order to simplify the analysis we assume that $\tilde{C}$ covers both messages and acknowledgements.)

12.4.1 The Upper Bound

In this section we prove an upper bound for the number T of rounds that is necessary to route all worms using the trial-and-failure protocol with some suitable values of Δ_t. We first present a structure that witnesses a long runtime of the protocol.

Assume that a worm w_0 is still active after t rounds. Then there must have been a worm w_1 that prevented it from moving forward in round t. But if w_0 and w_1 have been active at round t there must have been (not necessarily different) worms w_2 and w_3 which prevented w_0 and w_1 from moving forward in round $t - 1$. Continuing with this argumentation until round 1 we find:

If worm w_0 is still active after t rounds then a tree as given in Figure 12.3 can be constructed such that the nodes represent worms and two nodes with a common father a collision event.

Let us call this tree a *witness tree of depth t*, and denote it by $\mathcal{W}(t)$. The following definition formalizes what kind of embeddings of worms into the nodes of $\mathcal{W}(t)$ we only have to consider.

Definition 12.4.1. *Let φ be an embedding of worms into the nodes of $\mathcal{W}(t)$. A pair of worms (w, w') is called* collision pair *if $w \neq w'$, w is embedded in*

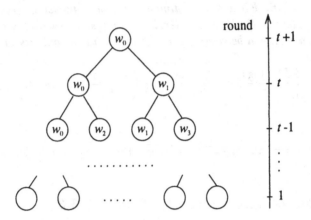

Fig. 12.3. The witness tree of depth t.

the left son, and w' is embedded in the right son of a common father in $\mathcal{W}(t)$. We call φ valid if for every collision pair (w, w') embedded at level i of $\mathcal{W}(t)$ it holds that

- w is also embedded in the father of w and w',
- there is no collision pair (w, w'') at level i with $w' \neq w''$, and
- the paths of w and w' share an edge.

A valid embedding is called active if for any collision pair (w, w') embedded at level i of $\mathcal{W}(t)$ it holds that w and w' use the same wavelength and w' prevents w from moving forward in round $t - i + 1$.

Following the discussion above, we can state the following lemma.

Lemma 12.4.1. If worm w_0 is still active after t rounds then there is an active embedding φ of worms into $\mathcal{W}(t)$ that maps w_0 to the root of $\mathcal{W}(t)$.

The above lemma implies that it suffices to find a suitable upper bound for the probability (w.r.t. random choices for the delays and wavelengths used by the worms) that there is an active embedding φ for any worm w_0 in order to prove the upper bound in Theorem 12.3.1.

In order to count the number of valid embeddings we introduce the following graph.

Definition 12.4.2. Let φ be a valid embedding. For each level $i \in \{1, \ldots, t\}$ of $\mathcal{W}(t)$, let $G_i = (V_i, E_i)$ be a directed graph whose nodes represent the set of worms embedded in level i and whose edges (w, w') represent the collision pairs (w, w') in level i. We call the worms in V_{i-1} old and the worms in $V_i \setminus V_{i-1}$ new w.r.t. G_i.

We assume G_0 to be the graph consisting only of a single node. Let the set of graphs $G_0, \ldots, G_t$ be called valid if they represent a valid embedding into $\mathcal{W}(t)$. Clearly, each valid embedding into $\mathcal{W}(t)$ has a unique valid set of graphs $G_0, \ldots, G_t$, and vice versa. Thus we can switch between either considering valid sets of graphs $G_0, \ldots, G_t$ or considering valid embeddings into $\mathcal{W}(t)$ in an arbitrary way.

For any valid embedding φ into the witness tree $\mathcal{W}(t)$, let $m_i = |V_i|$ denote the total number of worms and $\ell_i = m_i - m_{i-1}$ denote the number of new worms at level i. Let $\tilde{C}_j$ be an upper bound for the path congestion that holds at round j w.h.p. using the protocol above for suitably chosen $\Delta_1, \ldots, \Delta_j$ (determined later). Then it holds for the number $V(t, k)$ of valid embeddings in $\mathcal{W}(t)$ using k worms:

$$
V(t, k) \leq n \sum_{\substack{\ell_1, \ldots, \ell_t \geq 0, \\ \sum_i \ell_i = k-1}} \prod_{i=1}^{t} \binom{m_{i-1}}{\ell_i} \cdot \tilde{C}_{t-i+1}^{\ell_i} \cdot (\ell_i + m_{i-1})^{m_{i-1}-\ell_i} \, ,
$$

w.h.p. This formula is derived as follows.

- There are n ways to choose the worm that is embedded in the root of $\mathcal{W}(t)$.
- There are $\binom{m_{i-1}}{\ell_i}$ possibilities to choose ℓ_i old worms that collide with (and therefore narrow down the choices for) each of the ℓ_i new worms. Hence afterwards there are at most $\tilde{C}_{t-i+1}^{\ell_i}$ ways w.h.p. to choose the ℓ_i new worms.
- For the remaining $m_{i-1} - \ell_i$ old worms there are at most $\ell_i + m_{i-1}$ possibilities to choose the worm that prevents it from moving forward.

Before we can proceed with our calculation, we need an upper bound that holds for the path congestion after every round w.h.p., and need an upper bound for the probability that the embeddings counted in $V(t, k)$ are active.

Lemma 12.4.2. *For all $t \geq 2$ it holds that, if $\Delta_i \geq 8e\frac{L\tilde{C}}{B2^{i-1}}$ for all $i \in \{1, \ldots, t-1\}$, then the path congestion $\tilde{C}_t$ at round t is at most $\max\{\frac{\tilde{C}}{2^{t-1}}, O(\log n)\}$, w.h.p.*

Proof. The proof will be done by induction. Suppose, the path congestion at the beginning of round t is bounded by $\frac{\tilde{C}}{2^{t-1}} \geq 2\alpha \log n$ for some arbitrary constant $\alpha > 1$. Let $\Delta_t \geq 8e\frac{L\tilde{C}}{B2^{t-1}}$ be the delay range in round t. Consider any fixed worm w. Let $w_1, \ldots, w_k$ be the worms participating in round t whose paths share a link with the path of w, $k \leq \frac{\tilde{C}}{2^{t-1}}$. Further let the binary random variable $X_i = 1$ if and only if w_i fails to reach its destination in round t. Then $X = \sum_{i=1}^{k} X_i$ is a random variable denoting the path congestion of w after round t.

Since we only consider shortcut-free paths, it holds for every pair of worms w_i and w_j that

$$
\Pr[w_i \text{ is (partly) discarded by } w_j] \leq \frac{2L}{B\Delta_t} \, .
$$

Therefore,

$$\Pr[X_i = 1] \leq \frac{\tilde{C}_t \cdot 2L}{B\Delta_t} \leq \frac{1}{4e}$$

and hence $E[X] \leq \frac{\tilde{C}}{4e \cdot 2^{t-1}}$. Let $\mu = \frac{\tilde{C}}{4e \cdot 2^{t-1}}$. Since the worms choose their delays and wavelengths independently at random, we can use Chernoff bounds (see [HR90]) to prove that, for $\epsilon = 2e - 1$,

$$\Pr[X \geq (1+\epsilon)\mu] \leq \left(\frac{e}{1+\epsilon}\right)^{(1+\epsilon)\mu} = \left(\frac{1}{2}\right)^{2e\frac{\tilde{C}}{4e \cdot 2^{t-1}}} \leq \left(\frac{1}{2}\right)^{\alpha \log n} = \left(\frac{1}{n}\right)^{\alpha}.$$

For $\alpha > 1$, this yields the lemma. □

Hence in the following we assume that $\tilde{C}_i = \max\{\frac{\tilde{C}}{2^{i-1}}, O(\log n)\}$ for all $i \in \{1, \ldots, t\}$.

Next we bound the probability that any of the embeddings counted in $V(t, k)$ is active. As noted above, the probability that a collision pair (w, w') in level i of $\mathcal{W}(t)$ is active is at most $\frac{2L}{B\Delta_{t-i+1}}$. Let a node in G_i be called *root* if it has outdegree 0. Then we can prove the following nice property.

Lemma 12.4.3. *For every level i, the connected components in G_i are directed trees with new worms as roots.*

Proof. Every old worm needs a witness for its collision in round i and therefore can not be a root like the new worms, that have no witness since they are just introduced as witnesses in round i. Further a connected component can not have a cycle since

- in leveled path collections using the serve-first rule this would mean that worms prevent each other from moving forward. This however, is not possible in a leveled path collection.
- in shortcut-free path collections using the priority rule this would mean that a worm w_1 is discarded by a worm w_2 that has a higher priority than w_1, and w_2 is discarded by a worm w_3 that has a higher priority than w_2, and so on, until we arrive at a worm w_i that is discarded by w_1, since it has a higher priority than w_i. This, however, is not possible as long as no two worms with the same rank can meet in a round.

□

Since every directed tree in G_i of size s implies a probability of $\leq (\frac{2L}{B\Delta_{t-i+1}})^{s-1}$ that its edges correspond to collisions of worms, and since there are exactly ℓ_i trees in G_i, we obtain a probability of at most

$$\left(\frac{2L}{B\Delta_{t-i+1}}\right)^{(m_{i-1}+\ell_i)-\ell_i} = \left(\frac{2L}{B\Delta_{t-i+1}}\right)^{m_{i-1}}$$

that the collisions in level i are active. Therefore the probability $P(t, k)$ that there exists an active embedding in $\mathcal{W}(t)$ is at most

$$n \sum_{\substack{\ell_1,\dots,\ell_t \geq 0, \\ \sum_i \ell_i = k-1}} \prod_{i=1}^{t} \binom{m_{i-1}}{\ell_i} \cdot \tilde{C}_{t-i+1}^{\ell_i} \cdot (\ell_i + m_{i-1})^{m_{i-1}-\ell_i} \left(\frac{L}{B\Delta_{t-i+1}} \right)^{m_{i-1}}$$

In case that $\ell_i \leq m_{i-1}/2$, we get

$$\binom{m_{i-1}}{\ell_i}(\ell_i + m_{i-1})^{m_{i-1}-\ell_i} \leq \left(\frac{em_{i-1}}{m_{i-1}-\ell_i} \right)^{m_{i-1}-\ell_i} \left(\tfrac{3}{2}m_{i-1} \right)^{m_{i-1}-\ell_i}$$

$$\leq (3em_{i-1})^{m_{i-1}-\ell_i},$$

and otherwise (that is, $m_{i-1}/2 < \ell_i \leq m_{i-1}$)

$$\binom{m_{i-1}}{\ell_i}(\ell_i + m_{i-1})^{m_{i-1}-\ell_i} \leq 2^{2\ell_i}(2m_{i-1})^{m_{i-1}-\ell_i}.$$

Therefore,

$$P(t,k) \leq n \sum_{\substack{\ell_1,\dots,\ell_t \geq 0, \\ \sum_i \ell_i = k-1}} \prod_{i=1}^{t} 2^{2\ell_i} \cdot (3em_{i-1})^{m_{i-1}-\ell_i} \cdot \tilde{C}_{t-i+1}^{\ell_i} \left(\frac{2L}{B\Delta_{t-i+1}} \right)^{m_{i-1}}$$

$$\leq n \cdot \left(\frac{8L \cdot \tilde{C}}{B\Delta_1} \right)^{k-1} \sum_{\substack{\ell_1,\dots,\ell_t \geq 0, \\ \sum_i \ell_i = k-1}} \prod_{i=1}^{t} \left(\frac{6eLm_{i-1}}{B\Delta_{t-i+1}} \right)^{m_{i-1}-\ell_i}$$

if all Δ_i are chosen such that $\frac{\tilde{C}}{\Delta_1} \geq \frac{\tilde{C}_i}{\Delta_i}$. Furthermore, the following lemma holds:

Lemma 12.4.4. *If* $\Delta_i \geq \frac{40e^2 Lk}{B}$ *and* $\Delta_{i+1} \leq \Delta_i$ *for all* $i \in \{1,\dots,t-1\}$ *then*

$$\max_{\substack{\ell_1,\dots,\ell_t \geq 0, \\ \sum_i \ell_i = k-1}} \prod_{i=1}^{t} \left(\frac{6eLm_{i-1}}{B\Delta_{t-i+1}} \right)^{m_{i-1}-\ell_i} \leq \left(\frac{6eLt}{B\Delta_t} \right)^{\frac{1}{2}(t-\lceil \log k \rceil)^2}.$$

Proof. Above we saw that, for all $i \in \{1,\dots,t\}$, all connected components in G_i are trees in which the root always has to be a new worm. Hence from level $i-1$ to level i the number of worms has to increase by at least one. Thus $\sum_{i=1}^{t} m_i$ is minimized if we set $m_i = i+1$ for all $i \in \{1,\dots,t-\lceil \log k \rceil+1\}$, and $m_i = 2m_{i-1}$ for all $i > t - \lceil \log k \rceil + 1$. Then it holds that $\ell_i = 1$ for all $i \in \{1,\dots,t-\lceil \log k \rceil+1\}$, and $\ell_i = m_{i-1}$ for all $i > t - \lceil \log k \rceil + 1$. Therefore we get with $\Delta_{i+1} \leq \Delta_i$ for all $i \in \{1,\dots,t-1\}$:

$$\prod_{i=1}^{t} \left(\frac{6eLm_{i-1}}{B\Delta_{t-i+1}} \right)^{m_{i-1}-\ell_i} \leq \prod_{i=2}^{t-\lceil \log k \rceil+1} \left(\frac{6eLi}{B\Delta_{t-i+1}} \right)^{i-1}$$

$$\leq \left(\frac{6eLt}{B\Delta_t} \right)^{\sum_{i=1}^{t-\lceil \log k \rceil} i} \leq \left(\frac{6eLt}{B\Delta_t} \right)^{\frac{1}{2}(t-\lceil \log k \rceil)^2}.$$

Next we show that for all other distributions of the m_i it holds that

$$\prod_{i=1}^{t} \left(\frac{6eLm_{i-1}}{B\Delta_{t-i+1}}\right)^{m_{i-1}-\ell_i} \leq \left(\frac{6eLt}{B\Delta_t}\right)^{\frac{1}{2}(t-\lceil \log k\rceil)^2} \tag{12.1}$$

if all $\Delta_i \geq \frac{40e^2 Lk}{B}$.

Consider increasing the number m_j of worms at a stage $j < t$ with $m_j < m_{j+1}$ by 1. Then two terms in the product in (12.1) change: the $(i = j)$-term and the $(i = j + 1)$-term. Before increasing m_j, these terms are

$$\left(\frac{6eLm_{j-1}}{B\Delta_{t-j+1}}\right)^{m_{j-1}-\ell_j} \left(\frac{6eLm_j}{B\Delta_{t-j}}\right)^{m_j-\ell_{j+1}}, \tag{12.2}$$

and after increasing m_j by 1, they change to

$$\left(\frac{6eLm_{j-1}}{B\Delta_{t-j+1}}\right)^{m_{j-1}-(\ell_j+1)} \left(\frac{6eL(m_j+1)}{B\Delta_{t-j}}\right)^{(m_j+1)-(\ell_{j+1}-1)} \tag{12.3}$$

It holds $(12.2) \geq (12.3)$

$$\Leftrightarrow \quad \frac{6eLm_{j-1}}{B\Delta_{t-j+1}} \geq \left(\frac{m_j+1}{m_j}\right)^{m_j-\ell_{j+1}} \left(\frac{6eL(m_j+1)}{B\Delta_{t-j}}\right)^2$$

$$\Leftarrow \quad \frac{m_{j-1}}{\Delta_{t-j+1}} \geq e \cdot \frac{6eL(2m_{j-1}+1)^2}{B\Delta_{t-j}^2}$$

$$\Leftarrow \quad \frac{m_{j-1}}{\Delta_{t-j+1}} \geq e \cdot \frac{40eLm_{j-1}^2}{B\Delta_{t-j+1}^2}$$

$$\Leftarrow \quad \Delta_{t-j+1} \geq \frac{40e^2 Lk}{B}$$

Since any distribution of the m_i can be obtained from the initial distribution above by increasing one of the m_i by 1 again and again, the lemma follows. □

Clearly, there are $\binom{t+k-1}{t} \leq 2^{t+k-1}$ possibilities for choosing the $\ell_1, \ldots, \ell_t$ such that $\sum_{i=1}^{t} \ell_i = k - 1$. Thus we get for $\Delta_i \geq \frac{40e^2 Lk}{B}$ for all i that

$$P(t,k) \leq n \left(\frac{8L \cdot \tilde{C}}{B\Delta_1}\right)^{k-1} 2^{t+k-1} \left(\frac{6eLt}{B\Delta_t}\right)^{\frac{1}{2}(t-\lceil \log k\rceil)^2}$$

$$= n \cdot 2^t \left(\frac{16L \cdot \tilde{C}}{B\Delta_1}\right)^{k-1} \left(\frac{6eLt}{B\Delta_t}\right)^{\frac{1}{2}(t-\lceil \log k\rceil)^2}.$$

For any constant $\gamma > 0$, let

$$k_0 = \frac{(2+\gamma)\log n}{\log\left(2 + \frac{B}{16\tilde{C}}\left(\frac{D}{L}+1\right)\right)} + 1$$

and

$$T \geq \sqrt{\frac{2(2+\gamma)\log n}{\log\left(\left(\frac{\tilde{C}}{\log^{3/2} n} + \sqrt{\log n} + \frac{B}{\sqrt{\log n}}\left(\frac{D}{L}+1\right)\right)\right)}} + \lceil\log k_0\rceil .$$

If the routing takes more than T rounds then one of the following two cases must be true:

1. There must exist an active embedding into a witness tree $\mathcal{W}(t)$ with $t \leq T$ and $k \in \{k_0, \ldots, 2k_0\}$ different worms.
2. There must exist an active embedding into a witness tree $\mathcal{W}(T)$ with $k \leq k_0$ different worms.

Suppose that $\Delta_i \geq \max\{\frac{32L\cdot\tilde{C}_i}{B}, \frac{32L\cdot\tilde{C}}{B\log n}, \frac{40e^2 Lk_0}{B}\} + D + L$. Then we get:

Pr[The routing takes more than T rounds]

$\leq$ Pr[Case (1) holds] + Pr[Case (2) holds]

$$\leq \sum_{t=\log k_0}^{T} \sum_{k=k_0}^{2k_0} P(t,k) + \sum_{k=T}^{k_0} P(T,k)$$

$$\leq \sum_{t=\log k_0}^{T} \sum_{k=k_0}^{2k_0} n \cdot 2^t \left(\frac{16L\cdot\tilde{C}}{B\Delta_1}\right)^{k-1} +$$

$$\sum_{k=T}^{k_0} n \cdot 2^T \left(\frac{16L\cdot\tilde{C}}{B\Delta_1}\right)^{k-1} \left(\frac{6eLT}{B\Delta_T}\right)^{\frac{1}{2}(T-\lceil\log k\rceil)^2}$$

$$\leq \sum_{t=\log k_0}^{T} \sum_{k=k_0}^{2k_0} n \cdot 2^t \left(\frac{1}{2}\right)^{(2+\gamma)\log n} +$$

$$\sum_{k=T}^{k_0} n \cdot 2^T \left(\frac{1}{2}\right)^{k-1} \left(\frac{\sqrt{\log n}}{2e\left(\frac{\tilde{C}}{\log n} + \log n + B\left(\frac{D}{L}+1\right)\right)}\right)^{\frac{1}{2}(T-\lceil\log k\rceil)^2}$$

$$\leq \cdots \leq \frac{n^{-\gamma}}{2} + \frac{n^{-\gamma}}{2} \leq n^{-\gamma}$$

Therefore the overall runtime is

$$\sum_{t=1}^{T} (\Delta_t + 2(D+L))$$

$$= O\left(\sum_{t=1}^{T}\left(D + L + \frac{L}{B}\left(\frac{\tilde{C}}{2^{t-1}} + \frac{\tilde{C}}{\log n} + \log n\right)\right)\right),$$

w.h.p., which is bounded by

$$O\left(\frac{L \cdot \tilde{C}}{B} + \left(\sqrt{\log_\alpha n} + \log\log_\beta n\right)\left(\frac{L\log n}{B} + D + L\right)\right),$$

where $\alpha = \tilde{C} + B(\frac{D}{L} + 1) + 2$ and $\beta = 2 + \frac{B}{\tilde{C}}(\frac{D}{L} + 1)$. This completes the proof of the upper bound of Theorems 12.3.1 and 12.3.3.

12.4.2 The Lower Bound

In this section we will prove the lower bound in Theorems 12.3.1 and 12.3.3. We use a path collection that consists of the following two types of subcollections.

- Let $d = \lfloor\frac{L-1}{2}\rfloor + 1$. The first type consists of $n/(2\sqrt{\log n})$ structures consisting of $\sqrt{\log n}$ paths of length D that are connected as shown in Figure 12.4.

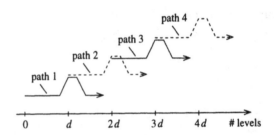

Fig. 12.4. A type-1 structure.

In general, the ith path starts in level $(i-1)d$ for all $i \geq 0$. Paths i and $i+1$ have a common edge from level $i \cdot d$ to level $i \cdot d + 1$.
- The second type consists of $n/(2\tilde{C})$ structures each consisting of $\tilde{C}$ identical paths of length D.

We assume that along each of these paths one worm of length $L \geq 2$ has to be sent.

We first compute how long it takes to route all worms in a type-1 structure. In case of routing along shortcut-free paths using priority routers, we assume that the worm traversing path i has rank i, and in case of conflicts worms with higher ranks are preferred. In order to bound the number of ways to assign delays and wavelengths to the worms such that conflicts occur, we need the following lemma.

Lemma 12.4.5. *Consider an arbitrary round of the trial-and-failure protocol with delay range $\Delta \geq L$. Suppose that the worms traversing the first $i+1$ paths are still active at the beginning of this round. Then with probability at least $\left(\frac{L-1}{2B\Delta}\right)^i$ the worms traversing the first i paths are discarded.*

Proof. Let Δ be the delay range of the round. Further let us denote the worm traversing path j by w_j, and its delay by δ_j. Clearly, there are $B\Delta$ ways to choose a wavelength and a delay for w_1. In the following we show that for any delay δ_i of worm w_i, there are at least $\frac{L-1}{2}$ ways to assign a delay δ_{i+1} to worm w_{i+1} such that w_{i+1} blocks w_i.

According to the construction of the type-1 structure, w_{i+1} starts $\lfloor\frac{L-1}{2}\rfloor +$ 1 levels after w_i. Hence, if $\delta_{i+1} \le \delta_i + \lfloor\frac{L-1}{2}\rfloor$ then w_{i+1} is at least one level ahead of w_i during the routing. On the other hand, if $\delta_{i+1} \ge \delta_i - \lfloor\frac{L-1}{2}\rfloor$ then w_{i+1} is at most $L-1$ levels ahead of w_i during the routing. Since $|\{\delta_i - \lfloor\frac{L-1}{2}\rfloor,\dots,\delta_i + \lfloor\frac{L-1}{2}\rfloor\} \cap [\Delta]| \ge \lfloor\frac{L-1}{2}\rfloor + 1 \ge \frac{L-1}{2}$ for $\Delta \ge L$, the number of ways to assign delays to w_{i+1} such that w_i is blocked by w_{i+1} is at least $\frac{L-1}{2}$.

Thus altogether there are at least $B\Delta(\frac{L-1}{2})^i$ ways to choose delays and wavelengths for the worms such that the worms traversing the first i paths are discarded. Hence this happens with a probability of at least

$$\frac{B\Delta(\frac{L-1}{2})^i}{(B\Delta)^{i+1}} = \left(\frac{L-1}{2B\Delta}\right)^i .$$

$\square$

Consider now the situation that it takes $t+1$ rounds to route the worms traversing the first $t+1$ paths in a type-1 structure. This could happen, e.g., if in round i only w_{t-i+2} is able to reach its destination, and the worms $w_1,\dots,w_{t-i+1}$ are discarded. According to the lemma above, for $L \ge 2$ the probability of such an event is at least

$$\prod_{i=1}^{t+1} \frac{B\Delta_i \left(\frac{L-1}{2}\right)^{t-i+1}}{(B(\Delta_i + L))^{t-i+2}} = \prod_{i=1}^{t} \left(\frac{L-1}{2B(\Delta_i + L)}\right)^{t-i+1}, \tag{12.4}$$

where $\Delta_i \ge 1$ is the delay range for round i. Clearly, the number of time steps necessary for the t rounds is at least $\Omega(\sum_{i=1}^{t}(\Delta_i + D + L))$. Given a fixed $\Delta = \sum_{i=1}^{t}\Delta_i$, the product in (12.4) gets minimal if $\Delta_i + L = (t-i+1)(\Delta + t\cdot L)/\binom{t+1}{2}$ for all $i \in \{1,\dots,t\}$. This is shown in the following lemma.

Lemma 12.4.6. *Consider* $x_1,\dots,x_n \in \mathbb{R}_+$ *with* $y = \sum_{i=1}^{n} x_i$. *Then, for every* $\alpha \in [0,y]$, $\prod_{i=1}^{n}(x_i + \alpha)^i$ *gets maximal if* $x_i + \alpha = i(y + n\cdot\alpha)/\binom{n+1}{2}$ *for all* $i \in \{1,\dots,n\}$.

Proof. For $n = 1$, the lemma is trivially true. We will show by complete induction on n that the assumption above is also true for $n > 1$. Suppose that we have already shown that $f(y,n) = \prod_{i=1}^{n}(i(y + n\alpha)/\binom{n+1}{2})^i$ is the maximal value the product of the $(x_i + \alpha)^i$ can reach. Then we want to find the $x_{n+1} + \alpha$ for which $f(y - x_{n+1},n) \cdot (x_{n+1} + \alpha)^{n+1}$ gets maximal. Clearly,

$$f(y - x_{n+1}, n) \cdot (x_{n+1} + \alpha)^{n+1} =$$
$$(y - x_{n+1} + n\alpha)^{\binom{n+1}{2}} \cdot (x_{n+1} + \alpha)^{n+1} \cdot g(n) , \tag{12.5}$$

where $g(n)$ is a function that only depends on n. Taking the logarithm yields

$$\log\left((y - x_{n+1} + n\alpha)^{\binom{n+1}{2}} \cdot (x_{n+1} + \alpha)^{n+1} \cdot g(n)\right)$$

$$= \binom{n+1}{2} \log(y - x_{n+1} + n\alpha) +$$

$$(n+1)\log(x_{n+1} + \alpha) + \log g(n) . \qquad (12.6)$$

Since a maximum of this function is also a maximum for the function in (12.5), it remains to determine the maximum of the function in (12.6). As (12.6) is a convex function, this can be done by finding the x_{n+1} for which the derivation of (12.6) is 0.

$$-\binom{n+1}{2} \cdot \frac{1}{y - x_{n+1} + n\alpha} + \frac{n+1}{x_{n+1} + \alpha} = 0$$

$$\Leftrightarrow \quad \frac{x_{n+1} + \alpha}{n+1} = \frac{y - x_{n+1} + n\alpha}{\binom{n+1}{2}}$$

$$\Leftrightarrow \quad x_{n+1} + \alpha = \frac{(n+1)(y + (n+1)\alpha)}{\binom{n+1}{2}} \cdot \frac{1}{1 + \frac{n+1}{\binom{n+1}{2}}}$$

$$\Leftrightarrow \quad x_{n+1} + \alpha = \frac{(n+1)(y + (n+1)\alpha)}{\binom{n+2}{2}}$$

Hence the x_i with $i \le n$ have to be chosen such that

$$\sum_{i}^{n}(x_i + \alpha) = y + (n+1)\alpha - \frac{(n+1)(y + (n+1)\alpha)}{\binom{n+2}{2}} .$$

According to the induction hypothesis, $\prod_{i=1}^{n}(x_i + \alpha)^i$ gets maximal if, for every $i \in \{1, \ldots, n\}$,

$$x_i + \alpha = \frac{i(y - x_{n+1} + n\alpha)}{\binom{n+1}{2}}$$

$$= \frac{i\left(y - \left(\frac{(n+1)(y+(n+1)\alpha)}{\binom{n+2}{2}} - \alpha\right) + n\alpha\right)}{\binom{n+1}{2}}$$

$$= \frac{i\left(\frac{\binom{n+2}{2}-(n+1)}{\binom{n+2}{2}} \cdot y + \frac{\binom{n+2}{2}-(n+1)}{\binom{n+2}{2}} \cdot (n+1)\alpha\right)}{\binom{n+1}{2}}$$

$$= \frac{i(y + (n+1)\alpha)}{\binom{n+2}{2}} .$$

From this the lemma follows. $\qquad\qquad\qquad\qquad\qquad\qquad\qquad\qquad \square$

Let $\bar{\Delta} = \Delta/t$. Since there are $n/(2\sqrt{\log n})$ type-1 structures, and each structure has a probability of at least

$$\prod_{i=1}^{t} \left(\frac{(L-1)(t+1)}{2B \cdot 2(t-i+1)(\bar{\Delta}+L)} \right)^{t-i+1} \geq \left(\frac{L-1}{4B(\bar{\Delta}+L)} \right)^{t^2}$$

to have active worms after t rounds, the expected number of type-1 structures that have active worms after t rounds is at least

$$\frac{n}{2\sqrt{\log n}} \left(\frac{L-1}{4B(\bar{\Delta}+L)} \right)^{t^2} < 1 \quad \Leftrightarrow \quad t \geq \sqrt{\frac{\log \left(\frac{n}{2\sqrt{\log n}} \right)}{\log \left(\frac{4B(\bar{\Delta}+L)}{L-1} \right)}}$$

Hence the expected number of rounds that are needed to route all worms in all type-1 structures is at least

$$\Omega \left(\sqrt{\log_{B(\bar{\Delta}/L+2)} n} \right) .$$

In order to bound the time needed to route all worms in the type-2 structures, we distinguish between the cases $\tilde{C} \geq 2^{\sqrt{\log n}}$ and $\tilde{C} \leq 2^{\sqrt{\log n}}$.

Case $\tilde{C} \leq 2^{\sqrt{\log n}}$:

Note that any routing protocol needs at least $\Omega(\frac{L\tilde{C}}{B} + D + L)$ steps to route all worms in a type-2 structure. Therefore the expected number of steps the protocol needs to route all worms is at least

$$\Omega \left(\frac{L\tilde{C}}{B} + \sqrt{\log_{B(\bar{\Delta}/L+2)} n}(\bar{\Delta} + D + L) \right) .$$

Since the runtime bound gets minimal for some $\bar{\Delta}$ chosen in $O(\frac{L\tilde{C}}{B} + D + L)$, the expected runtime of the protocol is at least

$$\Omega \left(\frac{L\tilde{C}}{B} + \sqrt{\log_\alpha n}(D + L) \right) ,$$

where $\alpha = \tilde{C} + B(\frac{D}{L} + 1) + 2$. Let $\beta = \alpha/\tilde{C} + 2$. Since $\tilde{C} \leq 2^{\sqrt{\log n}}$, it holds that $\sqrt{\log_\alpha n} \leq \log\log n$ only if $B(\frac{D}{L} + 1) \geq 2^{\log n/(\log\log n)^2} \gg \tilde{C}$. In this case, however, $\log \beta = \Theta(\log \alpha)$, that is, $\sqrt{\log_\alpha n} \geq \log\log_\beta n$. Therefore we arrive at an expected runtime of the protocol of at least

$$\Omega \left(\frac{L\tilde{C}}{B} + \left(\sqrt{\log_\alpha n} + \log\log_\beta n \right)(D + L) \right)$$

time steps.

Case $\tilde{C} \geq 2^{\sqrt{\log n}}$:

Let $\tilde{C}_i$ be the minimum over all type-2 structures P of the number of worms that are still active in P after i rounds. Then the following lemma holds.

Lemma 12.4.7. *For every $t \geq 2$ and $L(\frac{\tilde{C}}{B} + 2) \leq \Delta_1, \ldots, \Delta_{t-1} \leq \hat{\Delta}$ with $\tilde{C}/(\frac{32B\hat{\Delta}}{(L-1)\tilde{C}})^{2^{t-1}-1} \geq 7\ln n$ it holds that*

$$\tilde{C}_t \geq \frac{\tilde{C}}{\left(\frac{32B\hat{\Delta}}{(L-1)\tilde{C}}\right)^{2^{t-1}-1}}$$

w.h.p.

Proof. For $t = 1$, the bound on $\tilde{C}_t$ trivially holds. Suppose that the bound above for $\tilde{C}_t$ is true for some $t \geq 1$. Then we want to show that, if $\Delta_t \leq \hat{\Delta}$ and $\tilde{C}/(\frac{32B\hat{\Delta}}{(L-1)\tilde{C}})^{2^t-1} \geq 7\ln n$, then we get

$$\tilde{C}_{t+1} \geq \frac{\tilde{C}}{\left(\frac{32B\hat{\Delta}}{(L-1)\tilde{C}}\right)^{2^t-1}}$$

w.h.p. Assume in the following that $\tilde{C}/(\frac{32B\hat{\Delta}}{(L-1)\tilde{C}})^{2^t-1} \geq 6\lceil \alpha \ln n \rceil$ for some fixed constant $\alpha > 1$. Consider any fixed type-2 structure P. Let $w_1, \ldots, w_c$ be the worms participating in round t that use this type-2 structure, $c \geq \tilde{C}/(\frac{32B\hat{\Delta}}{(L-1)\tilde{C}})^{2^{t-1}-1}$. Further let the binary random variable $X_i = 1$ if and only if w_i fails to reach its destination in round t, and the binary random variable $D_i = 1$ if and only if w_i chooses a startup delay in $\{L, \ldots, \Delta_t - 1\}$ in round t. Then $X = \sum_{i=1}^c X_i$ is a random variable denoting the path congestion at P after round t. In order to bound $\sum_{i=1}^c X_i$, we need the following proposition.

Proposition 12.4.1. *For any $\{i_1, \ldots, i_\ell\} \subseteq \{1, \ldots, c/2\}$ with $\ell \leq 2\lceil \alpha \ln n \rceil$ it holds*

$$\Pr\left[\prod_{j=1}^\ell X_{i_j} = 1 \mid \prod_{j=1}^\ell D_{i_j} = 1\right] \geq \left(\frac{(L-1)\tilde{C}_t}{4B\hat{\Delta}}\right)^\ell.$$

Proof. In case of using priority routers, suppose that the priorities of the worms are chosen in such a way that $\text{rank}(w_1) < \text{rank}(w_2) < \ldots < \text{rank}(w_c)$, that is, w_c has the highest priority. Let $W = \{w_{c/2+1}, \ldots, w_c\}$. For all $i \in \{1, \ldots, c\}$, let E_i be the event that w_i fails because of a worm in W and F_i be the event that E_i is true under the assumption that $D_i = 1$. Then it holds for any subset $\{i_1, \ldots, i_\ell\} \subseteq \{1, \ldots, c/2\}$ that

$$\Pr\left[\prod_{j=1}^\ell X_{i_j} = 1 \mid \prod_{j=1}^\ell D_{i_j} = 1\right] \geq \Pr\left[\bigwedge_{j=1}^\ell F_{i_j}\right].$$

In order to proceed with our calculation, we need the following claim.

Claim 12.4.1.

$$\Pr[F_{i_1} \wedge \ldots \wedge F_{i_\ell}] \geq \prod_{j=1}^{\ell} \Pr[F_{i_j}] \ .$$

Proof. Let $\mathcal{A}$ be the set of all assignments $A \in ([B] \times [\Delta_t])^{c/2}$ of wavelengths and startup delays to the worms in W. Since the worms $w_{i_1}, \ldots, w_{i_\ell}$ choose wavelengths and startup delays independently at random, we get

$$\Pr[F_{i_\ell} \mid F_{i_1} \wedge \ldots \wedge F_{i_{\ell-1}}]$$

$$= \frac{\Pr[F_{i_1} \wedge \ldots \wedge F_{i_\ell}]}{\Pr[F_{i_1} \wedge \ldots \wedge F_{i_{\ell-1}}]} = \frac{\displaystyle\sum_{A \in \mathcal{A}} \Pr[F_{i_1} \wedge \ldots \wedge F_{i_\ell} \mid A]}{\displaystyle\sum_{A \in \mathcal{A}} \Pr[F_{i_1} \wedge \ldots \wedge F_{i_{\ell-1}} \mid A]}$$

$$= \frac{\displaystyle\sum_{A \in \mathcal{A}} \prod_{j=1}^{\ell} \Pr[F_{i_j} \mid A]}{\displaystyle\sum_{A \in \mathcal{A}} \prod_{j=1}^{\ell-1} \Pr[F_{i_j} \mid A]} = \frac{\displaystyle\sum_{A \in \mathcal{A}} (\Pr[F_{i_\ell} \mid A])^{\ell}}{\displaystyle\sum_{A \in \mathcal{A}} (\Pr[F_{i_\ell} \mid A])^{\ell-1}}$$

To proceed with our calculation we need the following inequality (see [Ho87], p. 226).

Claim 12.4.2 (Chebychev). *If* $a_1 \geq a_2 \geq \ldots \geq a_n \geq 0$ *and* $b_1 \geq b_2 \geq \ldots b_n \geq 0$ *then*

$$n \sum_{i=1}^{n} a_i b_i \geq \left(\sum_{i=1}^{n} a_i \right) \left(\sum_{i=1}^{n} b_i \right) \ .$$

Hence for every $\ell \geq 2$ and $x_1, \ldots, x_n \geq 0$ with $\sum_{i=1}^{n} x_i > 0$ it holds

$$\frac{\sum_{i=1}^{n} x_i^\ell}{\sum_{i=1}^{n} x_i^{\ell-1}} \geq \frac{1}{n} \sum_{i=1}^{n} x_i \ .$$

(Choose $a_i = x_i^{\ell-1}$ and $b_i = x_i$.) Thus we get

$$\Pr[F_{i_\ell} \mid F_{i_1} \wedge \ldots \wedge F_{i_{\ell-1}}] \geq \frac{1}{|\mathcal{A}|} \sum_{A \in \mathcal{A}} \Pr[F_{i_\ell} \mid A] = \Pr[F_{i_\ell}]$$

and therefore

$$\Pr[F_{i_1} \wedge \ldots \wedge F_{i_\ell}] \geq \prod_{j=1}^{\ell} \Pr[F_{i_j}] \ .$$

$\square$

It remains to bound $\Pr[F_i]$ for every $i \in \{1, \ldots, c/2\}$. Let $E_{f,\delta}$ denote the event that a worm in W chooses wavelength f and a startup delay in $[\delta - 1, \delta - (L - 1)]$. For every $i \in \{1, \ldots, c/2\}$, let f_i denote the wavelength and δ_i denote the startup delay chosen by worm w_i. Then we can show the following claim.

Claim 12.4.3. *If there are ℓ worms w_i, $1 \le i \le c/2$, for which E_{f_i, δ_i} is true, then at least ℓ worms are discarded.*

Proof. Let us choose any worm w in W that causes some, say a set W', of these ℓ worms w_i to have a true E_{f_i, δ_i}. Let δ be the delay of w. Then either all worms of W' are discarded because of w or w has been discarded and also at least $|W'| - 1$ other worms in W', since only one worm with a delay in $[\delta - 1, \delta - (L - 1)]$ can survive. In both cases at least $|W'|$ worms are eliminated. Summing over all worms in W yields the claim. $\square$

Thus we can assume that

$$\Pr[E_{f_i, \delta_i} \mid D_i = 1] \le \Pr[E_i \mid D_i = 1]$$

and therefore

$$
\begin{aligned}
\Pr[\neg E_i \mid D_i = 1] \;&\le\; \Pr[\neg E_{f_i, \delta_i} \mid D_i = 1] \\
&= \prod_{j=c/2+1}^{c} \left(1 - \Pr\left[\begin{array}{l} w_j \text{ uses } f_i \text{ and a delay in} \\ [\delta_i - 1, \delta_i - (L - 1)] \mid D_i = 1 \end{array} \right] \right) \\
&= \left(1 - \frac{L-1}{B\Delta_t} \right)^{c/2} .
\end{aligned}
$$

For $\Delta_t \ge \frac{L\tilde{C}}{B}$ we get

$$
\begin{aligned}
\Pr[F_i] \;&=\; \Pr[E_i \mid D_i = 1] \;\ge\; 1 - \left(1 - \frac{L-1}{B\Delta_t} \right)^{c/2} \\
&\ge\; 1 - \left(1 - \frac{c(L-1)}{4B\Delta_t} \right) = \frac{c(L-1)}{4B\Delta_t} \;\ge\; \frac{(L-1)\tilde{C}_t}{4B\Delta_t} .
\end{aligned}
$$

So altogether we get

$$\Pr\left[\prod_{j=1}^{\ell} X_{i_j} = 1 \mid \prod_{j=1}^{\ell} D_{i_j} = 1 \right] \ge \left(\frac{(L-1)\tilde{C}_t}{4B\Delta_t} \right)^{\ell} .$$

Since $\Delta_t \le \hat{\Delta}$, the proposition follows. $\square$

Let the random variable Y_i be defined as $Y_i = X_i \cdot D_i$. Proposition 12.4.1 implies that we can consider the random variables $Y_1, \ldots, Y_{c/2}$ as $(2\lceil \alpha \ln n \rceil)$-wise independent with probability at least

$$\frac{\Delta_t - L}{\Delta_t} \cdot \frac{(L-1)\tilde{C}_t}{4B\hat{\Delta}} \geq \frac{(L-1)\tilde{C}_t}{8B\hat{\Delta}}$$

if $\Delta_t \geq 2L$. In this case we get an expected path congestion after round t of

$$E[X] \geq \sum_{i=1}^{c/2} Y_i \geq \frac{\tilde{C}_t}{2} \cdot \frac{(L-1)\tilde{C}_t}{8B\hat{\Delta}} \ .$$

Let $\mu = \frac{\tilde{C}_t}{2} \cdot \frac{(L-1)\tilde{C}_t}{8B\hat{\Delta}}$. Then

$$\mu \geq \frac{L-1}{16B\hat{\Delta}} \cdot \left(\frac{\tilde{C}}{\left(\frac{32B\hat{\Delta}}{(L-1)\tilde{C}} \right)^{2^{t-1}-1}} \right)^2$$

$$= 2 \cdot \frac{\tilde{C}}{\left(\frac{32B\hat{\Delta}}{(L-1)\tilde{C}} \right)^{2^{t-1}-1+2^{t-1}-1+1}} = 2\frac{\tilde{C}}{\left(\frac{32B\hat{\Delta}}{(L-1)\tilde{C}} \right)^{2^t-1}}$$

In order to proceed with our calculation, we need the following result shown in [SSS95] (see Theorem 5).

Proposition 12.4.2. *Let X be a sum of k-wise independent binary random variables with $\mu \leq E[X]$, then it holds for $k \leq \lfloor \epsilon^2 \mu e^{-1/3} \rfloor$:*

$$\Pr[X \leq (1-\epsilon)\mu] \leq e^{-\lfloor k/2 \rfloor} \ .$$

Let $\epsilon = \frac{1}{2}$. Since $\tilde{C}/(\frac{32B\hat{\Delta}}{(L-1)\tilde{C}})^{2^{t-1}-1} \geq 6\lceil \alpha \ln n \rceil$, we have $2\lceil \alpha \ln n \rceil \leq \epsilon^2 \mu e^{-1/3}$. Thus we get together with Proposition 12.4.2:

$$\Pr\left[X \leq \frac{\tilde{C}}{\left(\frac{32B\hat{\Delta}}{(L-1)\tilde{C}} \right)^{2^t-1}} \right] \leq \Pr[X \leq (1-\epsilon)\mu]$$

$$\leq e^{-\alpha \ln n} = \left(\frac{1}{n} \right)^\alpha \ .$$

Hence for $\alpha > 1$ the path congestion after round t is at most $\tilde{C}/(\frac{32B\hat{\Delta}}{(L-1)\tilde{C}})^{2^t-1}$ for all type-2 structures, w.h.p.   □

Thus for any $L \geq 2$ and $\hat{\Delta} \geq 1$ it holds for the expected number t of rounds to route all worms in type-2 structures that

$$\frac{\tilde{C}}{\left(\frac{32B(\hat{\Delta}+L(\tilde{C}/B+2))}{(L-1)\tilde{C}} \right)^{2^t-1}} \leq 7 \ln n$$

$$\Leftrightarrow \quad t \geq \log\left(1 + \log_\gamma \frac{\tilde{C}}{7\ln n}\right),$$

where $\gamma = \frac{32B(\hat{A}+L(\tilde{C}/B+2))}{(L-1)\tilde{C}}$. Since $\tilde{C} \geq 2^{\sqrt{\log n}}$, the expected runtime of the protocol is at least

$$\Omega\left(\frac{L\tilde{C}}{B} + \left(\sqrt{\log_{\frac{B\hat{A}}{L}+2} n} + \log\log_\gamma n\right)(\bar{A} + D + L)\right).$$

Since this bound gets minimal for $\bar{A} = O(\frac{L\tilde{C}}{B} + D + L)$, we get an expected runtime of at least

$$\Omega\left(\frac{L\tilde{C}}{B} + \left(\sqrt{\log_\alpha n} + \log\log_\beta n\right)(D + L)\right)$$

time steps, where $\alpha = \tilde{C} + B(\frac{D}{L} + 1) + 2$ and $\beta = \alpha/\tilde{C} + 2$.

12.5 Proof of Theorem 12.3.2

In this section we prove upper and lower bounds on the runtime of our protocol for shortcut-free path collections using serve-first routers. Hence suppose we want to route worms of length L along a collection of n shortcut-free paths with path congestion $\tilde{C}$ and dilation D, using serve-first routers with bandwidth B. (We again assume that $\tilde{C}$ covers both messages and acknowledgments.)

12.5.1 The Upper Bound

In this section we prove the upper bound in Theorem 12.3.2. Let the witness tree $\mathcal{W}(t)$ be defined as in Section 12.4. For any valid embedding φ into $\mathcal{W}(t)$, let $m_i = |V_i|$ denote the total number of worms and $\ell_i = m_i - m_{i-1}$ denote the number of new worms at level i. Furthermore let c_i denote the number of old worms that are in a connected component in G_i with a new worm. Let $\tilde{C}_j$ be an upper bound for the path congestion that holds w.h.p. after round j using the trial-and-failure protocol for suitably chosen $\Delta_1, \ldots, \Delta_j$ (determined later). Then it holds for the number $V(t, k)$ of valid embeddings in $\mathcal{W}(t)$ using k worms:

$$V(t,k) \leq n \sum_{\substack{\ell_1, \ldots, \ell_t \geq 0, \\ \sum_i \ell_i = k-1}} \prod_{i=1}^{t} \sum_{c_i=\ell_i}^{m_{i-1}} \binom{m_{i-1}}{c_i}\binom{c_i}{\ell_i} \cdot$$
$$\tilde{C}_{t-i+1}^{\ell_i} \cdot (\ell_i + c_i)^{c_i-\ell_i} \cdot (m_{i-1} - c_i)^{m_{i-1}-c_i},$$

w.h.p. This formula is derived as follows.

- There are n ways to choose the worm that is embedded in the root of $\mathcal{W}(t)$.
- There are $\binom{m_{i-1}}{c_i}$ possibilities to choose c_i old worms that lie in a connected component in G_i with a new worm, and $\binom{c_i}{\ell_i}$ possibilities to choose ℓ_i old worms that collide with (and therefore narrow down the choices for) each of the ℓ_i new worms. Therefore afterwards there are at most $\tilde{C}_{t-i+1}^{\ell_i}$ ways w.h.p. to choose the ℓ_i new worms. For the remaining $c_i - \ell_i$ old worms there are at most $\ell_i + c_i$ possibilities to choose the worm that prevents it from moving forward.
- For each of the remaining $m_{i-1} - c_i$ old worms there are at most $m_{i-1} - c_i$ ways to determine the old worm which prevents it from moving forward.

Before we can proceed with our calculation, we need an upper bound that holds for the path congestion w.h.p., and need an upper bound for the probability that the embeddings counted in $V(t, k)$ are active.

Since the delays and wavelengths are chosen independently and we only consider shortcut-free paths, it holds for every pair of worms w_i and w_j at round t that

$$\Pr[w_i \text{ is blocked by } w_j] \leq \frac{L}{B\Delta_t} \ .$$

Therefore we get analogous to Lemma 12.4.2 that, if $\Delta_i \geq 4e\frac{L\tilde{C}}{B2^{i-1}}$ for all $i \in \{1,\ldots,t-1\}$, then the path congestion $\tilde{C}_t$ at round t is at most $\max\{\frac{\tilde{C}}{2^{i-1}}, O(\log n)\}$, w.h.p.

Next we bound the probability that the embeddings counted in $V(t, k)$ are active. As noted above, the probability that a collision pair (w, w') in level i of $\mathcal{W}(t)$ is active is at most $\frac{L}{B\Delta_{t-i+1}}$. For every level i, each connected component in G_i that contains no new worms has a size of at least three. This is true since we only allow the worms to be routed along shortcut-free paths and therefore two worms can not block each other. Hence there are at most $g_i \leq \frac{m_i - c_i}{3}$ components with no new worms. Since every connected component of size s implies a probability of at most $(\frac{L}{B\Delta_{t-i+1}})^{s-1}$ that its edges represent collisions of worms we obtain a probability of at most

$$\left(\frac{L}{B\Delta_{t-i+1}}\right)^{((m_{i-1}-c_i)-g_i)} \leq \left(\frac{L}{B\Delta_{t-i+1}}\right)^{\frac{2(m_{i-1}-c_i)}{3}}$$

that these components are active. Note that we can improve this bound if we know that at least $k > 3$ pieces of paths in the collection are needed to obtain a directed cycle in G_i.

According to Section 12.4, each connected component in G_i that contains a new worm forms a tree. Furthermore each new worm lies in a different connected component. Therefore the probability that their edges represent collisions of worms is at most

$$\left(\frac{L}{B\Delta_{t-i+1}}\right)^{(\ell_i+c_i)-\ell_i} \ .$$

Altogether the probability that all collision pairs in level i are active given m_{i-1} and c_i is at most

$$\left(\frac{L}{B\Delta_{t-i+1}}\right)^{c_i+\frac{2(m_{i-1}-c_i)}{3}} .$$

Therefore the probability $P(t,k)$ that there exists an active embedding in $\mathcal{W}(t)$ is at most

$$n \sum_{\substack{\ell_1,\ldots,\ell_t\geq 0, \\ \sum_i \ell_i = k-1}} \prod_{i=1}^{t} \sum_{c_i=\ell_i}^{m_{i-1}} \binom{m_{i-1}}{c_i}\binom{c_i}{\ell_i}\tilde{C}_{t-i+1}^{\ell_i}(\ell_i+c_i)^{c_i-\ell_i}.$$

$$(m_{i-1}-c_i)^{m_{i-1}-c_i}\left(\frac{L}{B\Delta_{t-i+1}}\right)^{c_i+\frac{2(m_{i-1}-c_i)}{3}} .$$

In order to simplify this formula, we have to distinguish between two cases. If $\ell_i \leq c_i/2$ we get

$$\binom{c_i}{\ell_i}(\ell_i+c_i)^{c_i-\ell_i} \leq \left(\frac{ec_i}{c_i-\ell_i}\right)^{c_i-\ell_i}\left(\tfrac{3}{2}c_i\right)^{c_i-\ell_i}$$
$$\leq (3ec_i)^{c_i-\ell_i} ,$$

and otherwise

$$\binom{c_i}{\ell_i}(\ell_i+c_i)^{c_i-\ell_i} \leq \tfrac{1}{2}2^{2\ell_i}(2c_i)^{c_i-\ell_i} .$$

Let $\Delta_i \geq \frac{216L}{B}$ for all $i \in \{1,\ldots,t\}$. Then it holds

$$\sum_{c_i=\ell_i}^{m_{i-1}} \binom{m_{i-1}}{c_i}\binom{c_i}{\ell_i}(\ell_i+c_i)^{c_i-\ell_i}(m_{i-1}-c_i)^{m_{i-1}-c_i}\left(\frac{L}{B\Delta_{t-i+1}}\right)^{c_i+\frac{2(m_{i-1}-c_i)}{3}}$$

$$\leq \sum_{c_i=\ell_i}^{m_{i-1}} \left(\frac{em_{i-1}}{m_{i-1}-c_i}\right)^{m_{i-1}-c_i}\binom{c_i}{\ell_i}(\ell_i+c_i)^{c_i-\ell_i}(m_{i-1}-c_i)^{m_{i-1}-c_i} .$$

$$\left(\frac{L}{B\Delta_{t-i+1}}\right)^{c_i+\frac{2(m_{i-1}-c_i)}{3}}$$

$$\leq (em_{i-1})^{m_{i-1}-\ell_i}\left(\frac{L}{B\Delta_{t-i+1}}\right)^{\ell_i+\frac{2(m_{i-1}-\ell_i)}{3}} .$$

$$\sum_{c_i=\ell_i}^{m_{i-1}} \binom{c_i}{\ell_i}(\ell_i+c_i)^{c_i-\ell_i}\left(\frac{1}{em_{i-1}}\right)^{c_i-\ell_i}\left(\frac{L}{B\Delta_{t-i+1}}\right)^{\frac{c_i-\ell_i}{3}}$$

$$\leq (em_{i-1})^{m_{i-1}-\ell_i}\left(\frac{L}{B\Delta_{t-i+1}}\right)^{\ell_i+\frac{2(m_{i-1}-\ell_i)}{3}} .$$

$$\sum_{c_i=\ell_i}^{m_{i-1}} \tfrac{1}{2} 2^{2\ell_i} (3ec_i)^{c_i-\ell_i} \left(\frac{1}{em_{i-1}}\right)^{c_i-\ell_i} \left(\frac{L}{B\Delta_{t-i+1}}\right)^{\frac{c_i-\ell_i}{3}}$$

$$\leq (em_{i-1})^{m_{i-1}-\ell_i} \left(\frac{L}{B\Delta_{t-i+1}}\right)^{\ell_i+\frac{2(m_{i-1}-\ell_i)}{3}} 4^{\ell_i} \cdot \tfrac{1}{2} \sum_{c_i=\ell_i}^{m_{i-1}} \left(\frac{27L}{B\Delta_{t-i+1}}\right)^{\frac{c_i-\ell_i}{3}}$$

$$\leq 4^{\ell_i} (em_{i-1})^{m_{i-1}-\ell_i} \left(\frac{L}{B\Delta_{t-i+1}}\right)^{\ell_i+\frac{2(m_{i-1}-\ell_i)}{3}}$$

Thus we get

$$P(t,k) \leq n \sum_{\substack{\ell_1,\ldots,\ell_t \geq 0,\\ \sum_i \ell_i = k-1}} \prod_{i=1}^{t} 4^{\ell_i} (em_{i-1})^{m_{i-1}-\ell_i} \tilde{C}_{t-i+1}^{\ell_i} \cdot$$

$$\left(\frac{L}{B\Delta_{t-i+1}}\right)^{\ell_i+\frac{2(m_{i-1}-\ell_i)}{3}}$$

$$\leq n \cdot \left(\frac{4L \cdot \tilde{C}}{B\Delta_1}\right)^{k-1} \sum_{\substack{\ell_1,\ldots,\ell_t \geq 0,\\ \sum_i \ell_i = k-1}} \prod_{i=1}^{t} \left(\frac{L(em_{i-1})^{3/2}}{B\Delta_{t-i+1}}\right)^{\frac{2(m_{i-1}-\ell_i)}{3}}$$

if all Δ_i are chosen such that $\frac{\tilde{C}}{\Delta_1} \geq \frac{\tilde{C}_i}{\Delta_i}$. Furthermore, the following lemma holds:

Lemma 12.5.1. *If $B\Delta_i \geq (3e)^3 Lk^{3/2}$ and $\Delta_{i+1} \leq \Delta_i$ for all $i \in \{1,\ldots,t-1\}$ then*

$$\max_{\substack{\ell_1,\ldots,\ell_t \geq 0,\\ \sum_i \ell_i = k-1}} \prod_{i=1}^{t} \left(\frac{L(em_{i-1})^{3/2}}{B\Delta_{t-i+1}}\right)^{\frac{2(m_{i-1}-\ell_i)}{3}} \leq \left(\frac{13L}{B\Delta_t}\right)^{t-\lceil \log k\rceil} .$$

Proof. We start with the valid embedding of k worms in $\mathcal{W}(t)$ that minimizes $\sum_{i=1}^{t} m_i$. Clearly, in each level $i \geq 1$ the number of different worms has to be at least 2. Hence we can choose the distribution $m_i = 2$ for all $i \in \{1,\ldots,t-\lceil \log k\rceil+1\}$. Then it holds that $\ell_1 = 1$ and $\ell_i = 0$ for all $i \in \{2,\ldots,t-\lceil \log k\rceil+1\}$. Therefore we get:

$$\prod_{i=1}^{t} \left(\frac{L(em_{i-1})^{3/2}}{B\Delta_{t-i+1}}\right)^{\frac{2(m_{i-1}-\ell_i)}{3}} \leq \prod_{i=2}^{t-\lceil \log k\rceil+1} \left(\frac{L(2e)^{3/2}}{B\Delta_{t-i+1}}\right)^{4/3}$$

$$\leq \left(\frac{13L}{B\Delta_t}\right)^{\frac{4}{3}(t-\lceil \log k\rceil)} .$$

In the following we show that for all other distributions of the m_i it holds that

$$\prod_{i=1}^{t} \left(\frac{L(em_{i-1})^{3/2}}{B\Delta_{t-i+1}} \right)^{\frac{2(m_{i-1}-\ell_i)}{3}} \leq \left(\frac{13L}{B\Delta_t} \right)^{\frac{4}{3}(t-\lceil \log k \rceil)} \tag{12.7}$$

if $B\Delta_i \geq (3e)^3 Lk^{3/2}$.

Consider increasing the number m_j of worms at a stage $j < t$ with $m_j < m_{j+1}$ by 1. Then two terms in the product in (12.7) change: the $(i = j)$-term and the $(i = j+1)$-term. Before increasing m_j, these terms are

$$\left(\frac{L(em_{j-1})^{3/2}}{B\Delta_{t-j+1}} \right)^{\frac{2(m_{j-1}-\ell_j)}{3}} \left(\frac{L(em_j)^{3/2}}{B\Delta_{t-j}} \right)^{\frac{2(m_j-\ell_{j+1})}{3}} \tag{12.8}$$

and after increasing m_j by 1, they change to

$$\left(\frac{L(em_{j-1})^{3/2}}{B\Delta_{t-j+1}} \right)^{\frac{2(m_{j-1}-(\ell_j+1))}{3}} \left(\frac{L(e(m_j+1))^{3/2}}{B\Delta_{t-j}} \right)^{\frac{2((m_j+1)-(\ell_{j+1}-1))}{3}} . \tag{12.9}$$

It holds

$$(12.8) \geq (12.9) \quad \Leftrightarrow \quad \left(\frac{L(em_{j-1})^{3/2}}{B\Delta_{t-j+1}} \right)^{2/3} \geq$$

$$\left(\frac{m_j+1}{m_j} \right)^{m_j - \ell_{j+1}} \left(\frac{L(e(m_j+1))^{3/2}}{B\Delta_{t-j}} \right)^{4/3}$$

$$\Leftarrow \quad \left(\frac{L(em_{j-1})^{3/2}}{B\Delta_{t-j+1}} \right)^{2/3} \geq e \cdot \left(\frac{L(e(2m_{j-1}+1))^{3/2}}{B\Delta_{t-j+1}} \right)^{4/3}$$

$$\Leftarrow \quad \frac{B\Delta_{t-j+1}}{Lm_{j-1}^{3/2}} \geq (3e)^3$$

$$\Leftarrow \quad B\Delta_{t-j+1} \geq (3e)^3 Lk^{3/2}$$

Since any distribution of m_i can be obtained from the initial distribution above by performing the action described above again and again, the lemma follows. □

Clearly, there are $\binom{t+k-1}{t} \leq 2^{t+k-1}$ possibilities for choosing the $\ell_1, \ldots, \ell_t$ such that $\sum_{i=1}^{t} \ell_i = k-1$. Thus we get for $B\Delta_i \geq (3e)^3 Lk^{3/2}$ for all i that

$$P(t,k) \leq n \left(\frac{4L \cdot \tilde{C}}{B\Delta_1} \right)^{k-1} 2^{t+k-1} \left(\frac{13L}{B\Delta_t} \right)^{t-\lceil \log k \rceil}$$

$$\leq n \cdot 2k \left(\frac{8L \cdot \tilde{C}}{B\Delta_1} \right)^{k-1} \left(\frac{26L}{B\Delta_t} \right)^{t-\lceil \log k \rceil} .$$

For any constant $\gamma > 0$, let

$$k_0 = \frac{(2+\gamma)\log n}{\log\left(2 + \frac{B}{8\tilde{C}}\left(\frac{D}{L}+1\right)\right)} + 1$$

and

$$T \geq \frac{(2+\gamma)\log n}{\log\left(\frac{\tilde{C}}{\log n} + \log^{3/2} n + \frac{B}{2}\left(\frac{D}{L}+1\right)\right)} + \lceil\log k_0\rceil .$$

If the routing takes more than T rounds then one of the following two cases must be true:

1. There must exist a valid reduced embedding into a witness tree $\mathcal{W}(t)$ with $t \leq T$ and $k \in \{k_0, \ldots, 2k_0\}$ different worms.
2. There must exist a valid reduced embedding into a witness tree $\mathcal{W}(T)$ with $k \leq k_0$ different worms.

Suppose that $\Delta_i \geq \max\{\frac{16L\cdot\tilde{C}_i}{B}, \frac{16L\cdot\tilde{C}}{B\log n}, \frac{(3e)^3 Lk_0^{3/2}}{B}\} + D + L$. Then we get:

$\Pr[\text{The routing takes more than } T \text{ rounds}]$

$\leq \Pr[\text{Case (1) holds}] + \Pr[\text{Case (2) holds}]$

$$\leq \sum_{t=\log k_0}^{T}\sum_{k=k_0}^{2k_0} P(t,k) + \sum_{k=2}^{k_0} P(T,k)$$

$$\leq \sum_{t=\log k_0}^{T}\sum_{k=k_0}^{2k_0} n\cdot 2k\left(\frac{8L\cdot\tilde{C}}{B\Delta_1}\right)^{k-1} +$$

$$\sum_{k=2}^{k_0} n\cdot 2k\left(\frac{8L\cdot\tilde{C}}{B\Delta_1}\right)^{k-1}\left(\frac{26L}{B\Delta_T}\right)^{T-\lceil\log k\rceil}$$

$$\leq \sum_{t=\log k_0}^{T}\sum_{k=k_0}^{2k_0} n\cdot 2k\left(\frac{1}{2}\right)^{(2+\gamma)\log n} +$$

$$\sum_{k=2}^{k_0} n\cdot 2k\left(\frac{1}{2}\right)^{k-1}\left(\frac{1}{2e\left(\frac{\tilde{C}}{\log n} + \log^{3/2} n\right) + \frac{B}{2}\left(\frac{D}{L}+1\right)}\right)^{T-\lceil\log k\rceil}$$

$$\leq \frac{n^{-\gamma}}{2} + \frac{n^{-\gamma}}{2} \leq n^{-\gamma}$$

Therefore the overall runtime is

$$\sum_{t=1}^{T}(\Delta_t + 2(D+L))$$

$$= O\left(\sum_{t=1}^{T}\left(D+L+\frac{L}{B}\left(\frac{\tilde{C}}{2^t} + \frac{\tilde{C}}{\log n} + \log^{3/2} n\right)\right)\right)$$

w.h.p., which is bounded by

$$O\left(\frac{L \cdot \tilde{C}}{B} + (\log_\alpha n + \log\log_\beta n)\left(\frac{L\log^{3/2} n}{B} + D + L\right)\right),$$

where $\alpha = \tilde{C} + B(\frac{D}{L} + 1) + 2$ and $\beta = \alpha/\tilde{C} + 2$.

12.5.2 The Lower Bound

In this section we will prove the lower bound in Theorem 12.3.2. We use a path collection that consists of the following two types of subcollections.

– The first type consists of $n/6$ structures consisting of three paths of length D that are connected as shown in Figure 12.5.

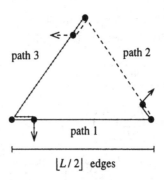

Fig. 12.5. A type-1 structure.

– The second type consists of $n/(2\tilde{C})$ structures each consisting of $\tilde{C}$ identical paths of length D.

We assume that along each of these paths one worm of length $L \geq 2$ has to be sent. (Note that in case of $L = 1$ no cycles of colliding worms can occur, that is, we are in a situation of Main Theorems 12.3.1 and 12.3.3.)

We first compute how long it takes to route all worms in a type-1 structure. Consider an arbitrary round i of the trial-and-failure protocol. Suppose that in a given type-1 structure all three worms are still active. Then we want to calculate the probability that these three worms also block each other in round i.

Suppose that $\Delta_i \geq L$. Let the worm traveling along path $j \in \{1, 2, 3\}$ be called w_j. Let δ_1 be the delay chosen by worm w_1. In case that $\delta_1 \leq \Delta_i - \lfloor \frac{L}{2} \rfloor$ we get that w_1, w_2, and w_3 collide if w_2 and w_3 choose the same wavelength as w_1 and delays in the range $[\delta_1, \delta_1 + \lfloor \frac{L}{2} \rfloor - 1]$. In case that $\delta_1 > \Delta_i - \lfloor \frac{L}{2} \rfloor$, then w_1, w_2, and w_3 collide if w_2 and w_3 choose the same wavelength as w_1

and delays in the range $[\delta_1 - (\lfloor \frac{L}{2} \rfloor - 1), \delta_1]$. Hence in both cases there are at least $\lfloor \frac{L}{2} \rfloor$ possibilities for both w_2 and w_3 to choose a wavelength and a delay such that w_1, w_2, and w_3 collide. Thus the probability that w_1, w_2, and w_3 collide at round i is at least $(\lfloor \frac{L}{2} \rfloor / (B\Delta_i))^2$ if $\Delta_i \geq L$. Therefore the probability that w_1, w_2, and w_3 collide for t rounds is at least

$$\prod_{i=1}^{t} \left(\frac{\lfloor L/2 \rfloor}{B(\Delta_i + L)} \right)^2$$

for any choice of $\Delta_1, \ldots, \Delta_t \geq 1$. Given a fixed $\Delta = \sum_{i=1}^{t} \Delta_i$ this product yields the smallest probability if $\Delta_i = \Delta/t$ for all $i \in \{1, \ldots, t\}$. Hence assume that all delay ranges are equal to $\bar{\Delta} = \Delta/t$. Since there are $n/6$ type-1 structures, and each structure has a probability of at least $(\frac{L}{3B(\bar{\Delta}+L)})^{2t}$ to have active worms after t rounds, the expected number of type-1 structures that have active worms after t rounds is at least

$$\frac{n}{6} \left(\frac{L}{3B(\bar{\Delta} + L)} \right)^{2t} < 1 \quad \Leftrightarrow \quad t \geq \frac{\log(n/6)}{2 \log \left(\frac{3B(\bar{\Delta}+L)}{L} \right)} .$$

Hence the expected number of rounds that are needed to route all worms is $\Omega(\log_{B(\bar{\Delta}/L+1)} n)$. In order to bound the time needed to route worms in the type-2 structures, we distinguish between the cases $\tilde{C} \geq 2^{\sqrt{\log n}}$ and $\tilde{C} \leq 2^{\sqrt{\log n}}$.

Case $\tilde{C} \leq 2^{\sqrt{\log n}}$:
Note that any routing protocol needs at least $\Omega(\frac{L\tilde{C}}{B} + D + L)$ steps to route all worms in a type-2 structure. Therefore the expected runtime of the protocol is at least

$$\Omega \left(\frac{L\tilde{C}}{B} + \log_{\frac{B\bar{\Delta}}{L}+2} n \cdot (\bar{\Delta} + D + L) \right)$$

$$= \Omega \left(\frac{L\tilde{C}}{B} + (\log_\alpha n + \log \log_\beta n)(D + L) \right),$$

where $\alpha = \tilde{C} + B(\frac{D}{L} + 1) + 2$ and $\beta = \alpha/\tilde{C} + 2$.

Case $\tilde{C} \geq 2^{\sqrt{\log n}}$:
This case follows analogous to Section 12.4.

13. Summary and Future Directions

At the end, let us summarize the main results in this book and present several open problems in the area of store-and-forward and wormhole routing. Furthermore, we will give an overview on topics that will become increasingly important in the future.

13.1 Store-and-Forward Routing

In the following we summarize the most important results that can be found in the store-and-forward part above and mention some important open problems. Let us first come back to the questions stated in the introduction. There we asked

- whether, and for which networks, online protocols can reach the best possible time of routing arbitrary permutations,
- whether, and for which networks, adaptive routing protocols are more efficient than protocols using oblivious routing strategies,
- whether, and for which networks, randomized routing strategies are more efficient than deterministic routing strategies, and
- how space limitations at the processors (such as bounded buffers or limited space for storing routing tables) influence the routing time.

Concerning the first question, we found out that, given any network of size N with routing number R at least $\Omega(\log^{1+\epsilon} N)$ for some constant $\epsilon > 0$, the online protocol by Ostrovsky and Rabani together with the strategies developed in Chapter 5 yield a protocol that can route any permutation in time $O(R)$, w.h.p., which is optimal in the worst case (and average case). For all bounded degree networks with smaller routing number, replacing the protocol by Ostrovsky and Rabani by the duplication protocol yields a protocol whose runtime deviates by at most a factor of $(\log \log N)^2$ from an optimal runtime of $O(R)$, w.h.p. It is still an open problem, whether also for these networks a runtime of $O(R)$ can be reached.

Concerning the second question, it follows from above that if random bits are available then for any network with sufficiently large routing number adaptive routing protocols can not beat protocols using oblivious routing strategies. This, however, only holds for the worst case (and average case) time of routing arbitrary permutations. We did not address the problem of reaching the optimal time for any *particular* permutation. The main obstacle in solving this problem is to construct for any permutation routing problem a path collection that is (asymptotically) optimal. If random bits are not available than oblivious protocols behave very poorly (see Theorem 5.2.1). Here we could show that adaptive protocols can reach much better time bounds (see Chapter 8). Note that for the case that only a limited randomness is used to choose paths in a given path system, Krizanc [Kr96] proved a general time-randomness tradeoff for the congestion of worst case permutations.

Concerning the third question, we could show in Chapter 8 that for any network with sufficiently large routing number there exists an (asymptotically) optimal deterministic routing strategy for routing arbitrary permutations using only constant size buffers. Hence for these networks and routing problems randomized strategies (asymptotically) can not beat deterministic strategies. For networks with small routing number this is still an open problem. Note that even for the butterfly network it is not known so far how fast deterministic strategies for routing arbitrary permutations can get.

We could also advance in answering the fourth question by presenting upper bounds in Chapter 9 for the relationship between space for storing routing information and the slowdown for routing arbitrary permutations in arbitrary networks.

In the following we summarize our results in more detail and give some more important open problems in the field of store-and-forward routing.

13.1.1 Path Selection

In Chapter 5 we could show that, for any network of size N with routing number $R = \Omega(\log N)$, it is possible to choose online for any permutation routing problem a path collection with a dilation and congestion of $O(R)$, which is asymptotically optimal in the worst case and even in the average case. However, there are permutations for which there are path collections with much less congestion and dilation (e.g., the permutation that maps x to x). Problems also arise if we want to find asymptotically optimal path collections for arbitrary functions and not only for permutations. Srinivasan and Teo [ST97] could show that for any routing problem in any network, an asymptotically optimal path collection can be found in polynomial time. But it is not known so far whether this can also be done online.

13.1.2 Offline Routing

We have seen that for any simple path collection with congestion C and dilation D there exists a protocol for routing a packet along each of these paths in optimal time $O(C+D)$, using buffers of size three. An open problem in this area is whether even for buffers of size one a runtime of $O(C + D)$

can be reached. Another interesting problem is, how small the constant in the $O(C + D)$ bound can be made if infinitely large buffers are available.

13.1.3 Oblivious Routing

In Chapter 7, we presented protocols that run efficiently for certain types of path collections. For instance, we showed that

- Ranade's protocol can route packets in optimal time, using only constant size buffers, along arbitrary leveled path collections of bounded degree with $\max\{C, D\}$ at least $\log n$, if all sources and all destinations resp. are in the same level,
- the growing rank protocol can route packets in optimal time along arbitrary shortcut-free path collections with $\max\{C, D\}$ at least $\log n$, and
- the protocol by Ostrovsky and Rabani can route packets in optimal time along arbitrary simple path collections with $\max\{C, D\}$ at least $\log^{1+\epsilon} n$ for some constant $\epsilon > 0$.

It is still an open problem, whether an optimal protocol for routing packets in arbitrary simple path collections with $\max\{C, D\}$ less than $\log^{1+\epsilon} n$ can also be constructed. Furthermore it is an open problem, how bounded buffers affect the routing time of oblivious routing strategies. The bounds we obtained so far (see Corollary 7.6.1 and Theorem 7.6.2) seem to be too weak. Observation 7.2.1 and Theorem 8.0.1 indicate that in leveled networks additional packets like ghost packets seem to be necessary for oblivious routing protocols to have an efficient routing time. A simpler strategy than Ranade's strategy could at least be found for butterfly networks (see [MS92]).

An interesting problem could also be to find an oblivious routing strategy that can route packets along some given path collection of size n in such a way that each packet that has to travel along a path of length d and has a maximum congestion of c at the edges of its path can be sent to its destination in $O(c + d + \log n)$ steps, w.h.p. Note that if all edges know their congestion in advance, this time bound can easily be reached for arbitrary shortcut-free path collections. Problematic seems to be the case that the edges do not know their congestion in advance.

Another open problem is how to route in arbitrary non-simple path collections. We know from the analysis of the extended growing rank protocol that this can be done efficiently if the path collection can be partitioned into subcollections of paths of approximately the same length and congestion that are shortcut-free. However, it is not known so far what to do if the lengths of the paths or congestion is very imbalanced within those subcollections, or too many stages of subcollections are necessary such that each one is shortcut-free. Interesting applications could be an efficient execution of several communication steps of a parallel program on a network.

Finally, it would be interesting to prove results about how ranks that are constructed using only a small seed of random bits influence the performance of routing protocols. Leighton *et al.* [LMRR94] showed that $\log n$-universal ranks suffice for Ranade's protocol to get the same time bound as shown in Theorem 7.3.1.

13.1.4 Adaptive Routing

We presented a deterministic routing strategy for routing arbitrary permutations in arbitrary networks in optimal time using constant size buffers if the routing number is large enough. Our results yield, for instance, that for every planar network of not too high degree there exists a deterministic protocol for routing any permutation in optimal time, using only constant size buffers.

One question that remains is, whether our upper bounds can be improved for networks with small routing number. Or, on the other hand, is there a non-trivial lower bound for deterministic routing on these networks? Since the constant hidden in the time bound of our strategy is large, it would furthermore be interesting to find deterministic algorithms that are not only asymptotically fast, but also applicable in practice.

As we noted in the previous section, it seems that oblivious routing protocols without using flow control strategies like sending ghost packets seem to perform poorly in leveled networks with bounded buffers. A protocol that does not need ghost packets could be the following: Use the random rank protocol. Each time a packet wants to use a buffer that is occupied by a packet with higher rank these packets are exchanged. This strategy might perform well, but no analysis for it is known so far, even for the butterfly.

For multibutterflies there are also still some interesting open problems left. Maggs and Vöcking [MV97] could recently show that arbitrary s-relations can be routed in optimal time even on bounded degree multibutterflies. However, is it possible to route also arbitrary functions in optimal time if combining is allowed? Note that Ranade could construct an optimal randomized routing strategy for this problem for the butterfly [Ra91].

13.1.5 Compact Routing

We started the chapter about compact routing by giving an overview about what is known for the relationship between space and stretch factor so far. Table 9.1 shows that for stretch factors smaller than two there are nodes that need $\Omega(N \log N)$ bits for storing routing tables, but for a stretch factor of at least 16 there are schemes that can do better. Therefore an important open question is, where in between stretch factor 2 and 16 we have for the first time the situation that all nodes need $o(N \log N)$ bits.

Furthermore, we presented a new approach for compact routing in arbitrary networks that relates space to the slowdown of routing strategies. We showed that for any bounded degree network of size N with sufficiently high routing number, $O(N^\delta)$ space suffices in the nodes for any constant $\delta > 0$

to construct a routing protocol that can route any permutation in optimal worst case time. Further we showed that for any bounded degree network of size N, $O(\log R \cdot \log \log R + \log N)$ space suffices in the nodes to route any permutation with slowdown at most $O(\log N / \log \log N)$.

The question is whether our results can be improved for arbitrary networks. Are there, for instance, other graph parameters apart from the routing number that can be used for more accurate results? In general, we would like to know what the precise upper and lower bounds are for compact routing. Our strategies for compact routing are asymptotically fast, but may perform poorly in practice. Are there other compact routing strategies that are more applicable in practice?

13.2 Wormhole Routing

In the following we summarize the most important results that can be found in the wormhole routing part above and mention some important open problems. Let us first come back to some of the questions stated in the introduction. There we asked

- how a limited bandwidth influences the routing time, and
- whether, and under which circumstances, wormhole routing strategies are more efficient than store-and-forward routing strategies for sending messages.

In order to address these questions, consider the following situation:

Along an arbitrary simple path collection of size n with congestion C and dilation D, messages have to be sent that consist of L flits each. For each flit, we need one time step to send it across a link.

There are two different ways of sending the packets to their destinations using store-and-forward routing:

1. If each message is considered as one packet, then it can be shown in a similar way as done in the proof of Theorem 6.2.1 that this requires time $\Theta(L(\frac{C}{B} + D))$.
2. If each message is broken into L independent packets, then it can be shown that this requires time $\Theta(L(\frac{C}{B} + 1) + D)$.

Using the wormhole routing strategy, this can be done online in

$$O\left(\frac{L \cdot C \cdot D^{1/B}}{B} + \left(\frac{\log n}{B} + 1\right)(D + L)\right)$$

steps, w.h.p., using the trial-and-failure protocol. Hence wormhole routing can beat strategy (1) for the case that, e.g., $B \geq \log D$ and $\log n \leq L \leq D$. Interestingly, wormhole routing can even reach the runtime of strategy (2) for the case that, e.g., $B \geq \log(nD)$. This demonstrates that keeping the flits of a message together in a worm-like fashion is not a restriction if the bandwidth is large enough.

In case that the bandwidth is less than $\log D$, the exact upper and lower bound for wormhole routing along arbitrary simple path collections seem to be close to

$$\frac{L \cdot C \cdot D^{1/B}}{B} + D + L \ .$$

For this it would be interesting to find a better offline protocol than that used for proving Theorem 10.2.1, and a more general lower bound than that presented in Theorem 10.2.2.

For bounded degree node-symmetric networks of size N with diameter D we showed that any permutation can be routed online in optimal time, w.h.p., if $B \geq \log D$ and $\log N \leq L \leq D$. An open problem is how to route in optimal time if $L \leq \log N$.

We further showed that for arbitrary d-dimensional meshes and tori with side length n any permutation can be routed online in optimal time, w.h.p., if $B \geq \log d$ and $L \cdot n \geq (d \cdot n + L) \log n$. Recently, Bock et al. [BMS97] showed that for $B = 1$ and any d and L such that $L = O(\frac{n}{(d \log n)^2})$, there is an online algorithm for the (n, d)-torus that routes any permutation of worms of length L in optimal time, w.h.p. An open problem is how to route in optimal time if B, d and L do not fit into the two cases above.

For the butterfly network we could show that even without using random delays and ranks, q worms of length L at each input can be routed online according to a randomly chosen function in optimal time, w.h.p., if $\log N \leq q \leq \log^\alpha N$ for any constant α, $L = \log N$, and $B \geq \log \log \log N$. An open problem is how to route in optimal time if, e.g., $L < \log N$ and/or $q < \log N$.

In case that wavelength conversion is not allowed we presented a very accurate analysis of the performance of (a variant of) the trial-and-failure protocol. The question is, what the real upper bound is for the runtime of the trial-and-failure protocol if wavelength conversion is allowed. (The bound presented in Theorem 11.1.1 seems to be too weak compared to the bounds obtained for all-optical routing.)

13.3 Future Directions

At the end of this book, let us give some future directions for the routing area and the techniques developed in the routing area.

13.3.1 Dynamic Routing

In contrast to static routing, in the dynamic routing model it is assumed that new messages are continually injected into the network. The advantage of this model is that it is much closer to real life than static routing. Early results in this area concentrated on strategies for the situation that some set of clients is connected via a bus to one server. The most popular protocol for this scenario is the *Binary Exponential Backoff Protocol* of Metcalfe and Boggs [MB76]. It is used for contention resolution in the Ethernet. In this protocol, a counter is associated to each request which keeps track of the number of times the request has failed to reach the server. In case that a request has already failed b times, it is retransmitted after t steps, where t is chosen uniformly at random from the set $\{1, \ldots, 2^b\}$ (in practice, t is chosen uniformly at random from $\{1, \ldots, 2^{\min\{10, b\}}\}$). Better protocols can be found in, e.g., [HLR87, RU95, PS95, GM96]. There has recently been a lot of progress also for dynamic routing in fixed connection networks such as butterflies and meshes. See e.g., [Mi94, HW95, KL95, BU96, DS97]. Universal dynamic routing protocols have also been presented [SV96].

Whereas all results above are based on a stochastic model of continuous message generation, recently a new model called *adversarial queueing theory* emerged. This approach was introduced by Borodin *et al.* in [BKR+96]. Their model permits an adversary to demand network bandwidth up to a prescribed *injection rate*. For $\epsilon > 0$, they say that an adversary *injects at rate* $1 - \epsilon$ if it generates requests (defined by arbitrary simple paths) subject to the following condition: For any edge e and during any t consecutive steps, the adversary can create at most $t(1 - \epsilon)$ requests that contain e in their path. Along each of these paths a packet has to be sent. For further results see [AAF96, AFH+97]. The advantage of the model of adversarial queueing theory over a stochastic message generation model is that the results are more robust in that they do not hinge upon particular probabilistic assumptions. Its disadvantage (at least for deterministic adversaries) is that the bounds on the delays of the packets are bounds for worst case scenarios. Therefore it can only be used to compare contention resolution rules under the worst case, which might rarely occur in real systems. So studying deterministic adversaries is useful if the stability of a system has to be maintained under any circumstance, whereas stochastic message generation models are useful for benchmarking systems.

There are still many open problems left in the area of dynamic routing strategies. For the stochastic message generation model, Scheideler and Vöcking [SV96] could present an optimal dynamic routing strategy for arbitrary shortcut-free path systems, using unbounded buffers. Their ideas can also be used together with Ranade's protocol to construct an asymptotically optimal dynamic routing strategy for arbitrary leveled path systems, using only edge buffers of constant size. However, an optimal dynamic routing strategy for arbitrary simple path systems is not known yet.

When using the stochastic message generation model, it is usually assumed that at each time step messages are generated independently of what happened at previous time steps. In practice, however, a high dependency to previous time steps can be observed especially in parallel systems. It is not clear so far how to model these dependencies, and whether there are universal strategies that are efficient for various models.

For the adversarial message generation model, it has been shown by Andrews *et al.* [AAF96] that there exist universal store-and-forward protocols that are stable for any constant injection rate $\lambda < 1$. The upper bounds they give for the delays of the packets, however, are at least D^2 (some even exponential in D), where D is the length of the longest path. An important open question is, whether this bound can be reduced to $O(D)$. For the special case of shortcut-free paths, this can be shown for injection rates $< 1/e$ by using a slight modification of the routing strategy in [SV96], if the adversary can not see the outcome of random experiments.

So far, mostly the situation has been considered of sending messages of uniform size. However, for multimedia applications this is not true any longer. The problem that arises here is first of all to find a realistic model for the distribution of the lengths of messages, the burstyness of the traffic, the duration of connections, quality of service requirements, and so on. In some scenarios even the loss of some packets may be allowed.

13.3.2 Routing in Faulty and Dynamic Networks

Another aspect we have not dealt with in this book is the possibility that nodes or links may develop faults or networks change dynamically.

Routing in Faulty Networks

Fault tolerance is a very important aspect for computer systems, since today the existence of whole companies or institutions can depend on whether their computer systems work reliably or not. There is a substantial body of literature concerning the fault-tolerance of communication networks. For an excellent survey see the paper by Leighton, Maggs and Sitaraman [LMS97]. Usually, people in this area are concerned with how long a network remains connected, or how much a collection of faults can slow down some computation in the network. An important topic is, for instance, to study the length of time it takes an impaired network to emulate a single step of a fault-free network of the same type and size. Of particular interest is to prove bounds on the number of faults that can be tolerated without losing more than a constant factor in speed. Strategies that are typically used are to embed a fault-free network into a faulty network that avoids the faults, or to use redundant computation. Even for some popular networks like the butterfly there are still interesting problems left. For instance, can a butterfly tolerate random-faults with constant failure probability or more than $N^{1-\epsilon}$ worst-case faults and still emulate a fault-free butterfly of the same size with constant slowdown?

With the rise of new computational models like the BSP model [Va90], algorithms are developed that can be efficiently executed on a wide range of parallel systems. That is, these algorithms do not require or assume a specific network topology any more. All they require is that certain high-level routing functions like permutation routing or broadcasting can be completed within some given period of time. This allows the operating system to be much more flexible, since it is not important any more for an impaired network to emulate a single step of a fault-free network of the *same* type and size. Of great interest would be to prove general results about the fault tolerance of networks by using parameters like the expansion or routing number of a network, or to find parameters that are more suitable.

Routing in Dynamic Networks

Only a few papers have studied dynamic networks so far. For a survey, see the paper by Dolev *et al.* [DKKP95]. Mostly, the problem of maintaining an efficient path system in a dynamically changing network is studied. The efficiency is usually measured by using the stretch factor (see Section 3.4.5 for a definition). So the problem is to find a strategy that can guarantee for some certain rate of topology changes a stretch factor that is as low as possible.

In case that a network is only lightly loaded with messages, the stretch factor is a very accurate parameter for measuring the quality of path systems. This, however, is only true for lightly loaded networks, since the stretch factor does not give any information about how many paths share the same link. So for highly loaded networks the congestion can be catastrophic even for path systems with constant stretch factor. In this case it would be therefore more interesting to find strategies that can maintain for some certain rate of topology changes a path system that has low dilation and expected congestion. As shown in Chapter 5, such strategies could then be used to ensure a fast online routing of arbitrary permutations.

Despite the fact that only a few people have studied dynamic networks so far, this model and its possible extensions have the potential to become very important in the future. The reason for this is that at the moment a dramatic growth in wireless networks can be observed. Mobile hosts and wireless networking hardware are becoming widely available, and extensive work has been done recently in integrating these elements into traditional networks such as the Internet. There are, however, important scenarios in which no fixed wired infrastructure such as the Internet is available, either because it may not be economically practical or physically possible to provide the necessary infrastructure or because the expediency of the situation does not permit its installation. For example, a class of students may need to interact during a lecture, friends or business associates may run into each other in an airport terminal and wish to share files, or a group of emergency rescue workers may need to be quickly deployed after an earthquake or flood. In such situations, a collection of mobile hosts with wireless network interfaces may form a temporary network without the aid of any established infrastructure or centralized administration. This type of wireless network is known as *ad hoc* network (see, e.g., [Gi96, IK96]).

If only two hosts, located closely together, are involved in the ad hoc network, no real routing protocol or routing decisions are necessary. In many ad hoc networks, though, two hosts that want to communicate may not be within transmission range of each other or may have to block too many other hosts in case of a direct transmission. In this case it might be useful or even necessary to forward a message along a sequence of hosts in the network (i.e., to perform several *hops*). Apart from the problem of finding the right route and scheduling the packets of different transmissions, the movements of the mobile hosts force the network to update its route system from time to time. Hence we have the problem of routing in (extensions of) dynamic networks. We believe that solving these problems will become very important in the future.

13.3.3 Scheduling

Another possible direction is to look at generalizations of routing problems, such as scheduling problems. There is a wide range of scheduling models. One example is shop scheduling (see Section 1.4.2). Shop scheduling models are defined in terms of resources, called *machines*, and activities, or *jobs*, each of which consists of a set of *operations*. Each operation of a job has an associated machine that it must be processed on for a given length of time. Each machine can process at most one operation at a time, and each job can have at most one of its operations undergoing processing at any point in time. The problem is to deliver the jobs among the machines in such a way that the time to complete all jobs becomes minimal.

Shop scheduling has many interesting applications such as to optimize production processes. As can be seen in Section 1.4.2, many problems in this area are still open. Since shop scheduling can be viewed as a generalization of packet scheduling, it might be possible to use techniques developed for analyzing routing strategies to solve shop scheduling problems.

There are at least three important generalizations of the job shop problem. The first is *dag shop scheduling*, where each job consists of a set of operations on different machines that must be processed in an order consistent with a given partial order. (For job shop scheduling, this partial order is always a chain, while for flow shop the partial order is the same chain for all jobs.) Note that also here we still require that no two operations of the same job can be processed simultaneously. One can further generalize the problem to the situation where, rather than having m different machines, there are m' machines, and for each type, there are a specified number of identical machines. Each operation, rather than being assigned to one machine, may be processed on any machine of the appropriate type. These problems have significant practical importance, since in real-world shops we would expect that a job need not to follow a strict total order and that the shop would have more than one copy of many of their machines. A further generalization is to allow different operations of the same job to be processed simultaneously, as long as this does not violate the partial ordering of the operations. To our knowledge, only results are known for these generalizations that are direct adaptations of results in the field of job shop scheduling. Much more work is definitely needed here.

References

[AAF96] M. Andrews, B. Awerbuch, A. Fernández, J. Kleinberg, T. Leighton, Z. Liu. Universal stability for greedy contention-resolution protocols. In *Proc. of the 35th IEEE Symposium on Foundations of Computer Science*, pp. 380-389, 1996.

[AAMR93] W. Aiello, B. Awerbuch, B. Maggs, S. Rao. Approximate load balancing on dynamic and asynchronous networks. In *Proc. of the 25th Ann. ACM Symp. on Theory of Computing*, pp. 632-641, 1993.

[AFH+97] M. Andrews, A. Fernández, M. Harcol-Balter, T. Leighton, L. Zhang. General dynamic routing with per-packet delay guarantees of O(distance + 1/session rate). In *Proc. of the 36th Ann. IEEE Symp. on Foundations of Computer Science*, pp. 294-302, 1997.

[ABC94] A. Aggarwal, A. Bar-Noy, D. Coppersmith, R. Ramaswami, B. Schieber, M. Sudan. Efficient routing and scheduling algorithms for optical networks. In *Proc. of the 5th Ann. ACM-SIAM Symp. on Discrete Algorithms*, pp. 412-423, 1994.

[ABLP90] B. Awerbuch, A. Bar-Noy, N. Linial, D. Peleg. Improved routing strategies with succinct tables. *Journal of Algorithms* **11**, pp. 307-341, 1990.

[ACG94] N. Alon, F.R.K. Chung, R.L. Graham. Routing permutations on graphs via matchings. *SIAM J. on Discrete Mathematics* **7**(3), pp. 513-530, 1994.

[AES92] N. Alon, P. Erdős, J. Spencer. *The Probabilistic Method*. Wiley Interscience Series in Discrete Mathematics and Optimization, John Wiley & Sons, 1992.

[AHMP87] H. Alt, T. Hagerup, K. Mehlhorn, F.P. Preparata. Deterministic simulation of idealized parallel computers on more realistic ones. *SIAM J. on Computing* **16**(5), pp. 808-835, 1987.

[AHS91] J. Aspnes, M. Herlihy, N. Shavit. Counting networks and multiprocessor co-ordination. In *Proc. of the 23rd Ann. ACM Symp. on Theory of Computing*, pp. 348-358, 1991.

[AISS95] A. Alexandrov, M.F. Ionescu, K.E. Schauser, C. Scheiman. LogGP: Incorporating long messages into the LogP model. In *Proc. of the 7th Ann. ACM Symp. on Parallel Algorithms and Architectures*, pp. 95-105, 1995.

[AKS83] M. Ajtai, J. Komlós, E. Szemerédi. Sorting in $c \log n$ parallel steps. *Combinatorica* **3**, pp. 1-19, 1983.

[Al82] R. Aleliunas. Randomized parallel communication. In *Proc. of the 1st Ann. ACM Symp. on Principles of Distributed Computing*, pp. 60-72, 1982.

[ALM96] S. Arora, F.T. Leighton, B.M. Maggs. On-line algorithms for path selection in a nonblocking network. *SIAM J. Comput.* **25**(3), pp. 600-625, 1996.

[AP92] B. Awerbuch, D. Peleg. Routing with polynomial communication-space tradeoff. *SIAM J. on Discrete Mathematics* **5**, pp. 151-162, 1992.

[AS92] A. Acampora, S. Shah. Multihop lightwave networks: A comparison of store-and-forward and hot-potato routing. *IEEE Trans. on Communications* **40**, pp. 1082-1090, 1992.

[AV79] D. Angluin, L.G. Valiant. Fast probabilistic algorithms for Hamiltonian circuits and matchings. *J. of Computer and System Sciences* **18**, pp. 155-193, 1979.

[Ba64] P. Baran. On distributed communication networks. *IEEE Trans. on Communications*, pp. 1-9, 1964.

[Ba68] K. Batcher. Sorting networks and their applications. In *Proc. of the AFIPS Spring Joint Computing Conference* **32**, pp. 307-314, 1968.

[BCFR96] T.A. Birks, D.O. Culverhouse, S.G. Farwell, P.St.J. Russel. 2×2 single-mode fiber routing switch. *Optics Letters* **21**, pp. 722-724, 1996.

[BCC+88] S. Borkar, R. Cohn, G. Cox, S. Gleason, T. Gross, H.T. Kung, M. Lam, B. Moore, C. Peterson, J. Pieper, L. Rankin, P.S. Tseng, J. Sutton, J. Urbanski, J. Webb. iWarp, an integrated solution to high-speed parallel computing. In *Proc. of the 1988 Int. Conf. on Supercomputing*, pp. 330-339, 1988.

[BCH+88] S.N. Bhatt, F.R.K. Chung, J.-W. Hong, F.T. Leighton, A.L. Rosenberg. Optimal simulations by butterfly networks. In *Proc. of the 20th Ann. ACM Symp. on Theory of Computing*, pp. 192-204, 1988.

[BDM95] A. Bäumker, W. Dittrich, F. Meyer auf der Heide. Truly efficient parallel algorithms: c-optimal multisearch for an extension of the BSP model. In *Proc of the 3rd Ann. European Symp. on Algorithms*, pp. 17-30, 1995.

[Be65] V.E. Beneš. *Mathematical Theory of Connecting Networks and Telephone Traffic*. Academic Press, New York, 1965.

[Be91] J. Beck. An algorithmic approach to the Lovász Local Lemma. *Random Structures and Algorithms* **2**(4), pp. 367-378, 1991.

[BFSU92] A. Broder, A. Frieze, E. Shamir, E. Upfal. Near-perfect token distribution. In *Proc. of the 19th Int. Colloquium on Automata, Languages and Programming*, pp. 308-317, 1992.

[BH85] A. Borodin, J.E. Hopcroft. Routing, merging, and sorting on parallel models of computation. *J. of Computer and System Sciences* **30**, pp. 130-145, 1985.

[BH92] R.A. Barry, P.A. Humblet. Bounds on the number of wavelengths needed in WDM networks. In *LEOS'92 Summer Topical Mtg. Digest*, pp. 114-127, 1992.

[BH94] R.A. Barry, P.A. Humblet. On the number of wavelengths and switches in all-optical networks. In *IEEE Trans. on Communications* **42**, pp. 583-591, 1994.

[BKR+96] A. Borodin, J. Kleinberg, P. Raghavan, M. Sudan, D.P. Williamson. Adversarial queueing theory. In *Proc. of the 28th ACM Symposium on Theory of Computing*, pp. 376-385, 1996.

[Bl95] U. Black. *ATM: Foundations for Broadband Networks*. Prentice Hall, 1995.

[BMS97] S. Bock, F. Meyer auf der Heide, C. Scheideler. Optimal wormhole routing in the (n,d)-torus. In *IPPS 97*.

[Bo96] S. Bock. Optimales Wormhole Routing im hochdimensionalen Torus. Diploma thesis, Paderborn University, March 1996.

[Br90] C. Brackett. Dense wavelength division multiplexing networks: Principles and applications. *IEEE J. Selected Areas in Comm.* **8**, pp. 373-380, August 1990.

[BRST93] A. Bar-Noy, P. Raghavan, B. Schieber, H. Tamaki. Fast deflection routing for packets and worms. In *Proc. of the 12th Ann. ACM Symp. on Principles of Distributed Computing*, pp. 75-86, 1993.

[BRSU93] A. Borodin, P. Raghavan, B. Schieber, E. Upfal. How much can hardware help routing? In *Proc. of the 25th Ann. ACM Symposium on Theory of Computing*, pp. 573-582, 1993.

[BS91] K. Bala, T.E. Stern. Algorithms for routing in a linear lightwave network. In *Proc. of INFOCOM*, pp. 1-9, 1991.

[BS94] I. Ben-Aroya, A. Schuster. Greedy hot-potato routing on the two-dimensional mesh. In *Proc. of the 2nd European Symp. on Algorithms*, 1994.

[BT89] D.P. Bertsekas, J.N. Tsitsiklis. *Parallel and Distributed Computation: Numerical Methods*. Chapter 7, Prentice-Hall, Englewood Cliffs, NJ, 1989.

[BU96] A. Broder, E. Upfal. Dynamic deflection routing on arrays. In *Proc. of the 28th ACM Symposium on Theory of Computing*, pp. 348-355, 1996.

[CG94] R. Cypher, L. Gravano. Storage-efficient, deadlock-free, adaptive packet routing algorithms for torus networks. *IEEE Trans. on Computers* 43(12), pp. 1376-1385, 1994.

[CGK92] I. Chlamtac, A. Ganz, G. Karmi. Lightpath communications: A novel approach to high bandwidth optical WANs. *IEEE Trans. on Communications* 40, July 1992.

[CGK93] I. Chlamtac, A. Ganz, G. Karmi. Lightnets: Topologies for high-speed optical networks. *IEEE/OSA J. Lightwave Techn.* 11(5/6), pp. 951-961, 1993.

[Ch94] P.J. Chidgey. Multi-wavelength transport networks. *IEEE Communications Magazine*, Dez. 1994, pp. 28-35.

[Ch95] V.W.S. Chan. All-optical networks. *Scientific American*, Sept. 1995, pp. 56-59.

[CK80] Y.C. Chow, W. Kohler. Models for dynamic load balancing in a heterogeneous multiple processor system. *IEEE Trans. on Computers* **C-28**(5), pp. 57-68, 1980.

[CLT96] D.D. Chinn, T. Leighton, M. Tompa. Minimal adaptive routing on the mesh with bounded queue size. *J. of Parallel and Distributed Computing* **34**, pp. 154-170, 1996.

[CMS95] A. Czumaj, F. Meyer auf der Heide, V. Stemann. Shared memory simulations with triple-logarithmic delay. In *3rd European Symp. on Algorithms*, pp. 46-59, 1995.

[CMS96] R.J. Cole, B.M. Maggs, R.K. Sitaraman. On the benefit of supporting virtual channels in wormhole routers. In *Proc. of the 8th Ann. ACM Symp. on Parallel Algorithms and Architectures*, pp. 131-141, 1996.

[CMSV96] R. Cypher, F. Meyer auf der Heide, C. Scheideler, B. Vöcking. Universal algorithms for store-and-forward and wormhole routing. In *28th Ann. ACM Symp. on Theory of Computing*, pp. 356-365, 1996.

[CNW90] N.K. Chung, K. Nosu, G. Winzer. *IEEE JSAC: Special Issue on Dense WDM Networks* 8, 1990.

[CP93] R. Cypher, C.G. Plaxton. Deterministic Sorting in nearly logarithmic time on the hypercube and related computers. *J. of Computer and System Sciences* **47**, pp. 501-548, 1993.

[Cy89] G. Cybenko. Dynamic load balancing for distributed memory multiprocessors. *J. of Parallel and Distributed Computing* **2**(7), pp. 279-301, 1989.

[Cy95] R. Cypher. Minimal deadlock-free routing in hypercubic and arbitrary networks. In *Proc. of the 7th IEEE Symp. on Parallel and Distributed Processing*, pp. 122-129, 1995.

[Da90] W. Dally. Virtual-channel flow control. In *Proc. of the 17th Int. Symp. on Computer Architecture*, pp. 60-68, 1990.

[Da92] W. Dally. Virtual-channel flow control. *IEEE Trans. on Parallel and Distributed Systems* 3(2), pp. 194-205, 1992.

[Di91] M. Dietzfelbinger. *Universal Hashing in Sequential, Parallel, and Distributed Computation*. Habilitation Thesis, Paderborn University, September 1991.

[Di96] M. Di Ianni. Efficient delay routing. In *2nd Int. Euro-Par Conference*, Vol. I, pp. 258-269, 1996.

[DKKP95] S. Dolev, E. Kranakis, D. Krizanc, D. Peleg. Bubbles: Adaptive routing scheme for high-speed dynamic networks. In *Proc. of the 27th Ann. ACM Symp. on Theory of Computing*, pp. 528-537, 1995.

[DS87] W. Dally, C. Seitz. Deadlock-free message routing in multiprocessor interconnection networks. *IEEE Trans. on Computers* **C-36**(5), pp. 547-553, 1987.

[DS97] S. Datta, R. Sitaraman. The performance of simple routing algorithms that drop packets. To appear in *Proc. of the 9th ACM Symposium on Parallel Algorithms and Architectures*, pp. 159-169, 1997.

[DV93] D.H.C. Du, R.J. Vetter. Distributed computing with high-speed optical networks. *IEEE Computer* **26**, pp. 8-18, 1993.

[ELZ86] D. Eager, E. Lazowska, J. Zahorjan. Adaptive load sharing in homogeneous distributed systems. *IEEE Trans. on Software* **12**(5), pp. 662-675, 1986.

[FG94] P. Fraigniaud, C. Gavoille. Optimal interval routing. In *CONPAR '94 – VAPP VI*, Springer Verlag, LNCS 854, pp. 785-796, 1994.

[FG95] P. Fraigniaud, C. Gavoille. Local memory requirement for universal routing schemes. In *14th Ann. ACM Symp. on Principles of Distributed Computing*, pp. 223-230, 1995.

[FG96] P. Fraigniaud, C. Gavoille. Local memory requirement for universal routing schemes. In *8th Ann. ACM Symp. on Parallel Algorithms and Architectures*, pp. 183-188, 1996.

[FGS93] M. Flammini, G. Gambosi, S. Salomone. Boolean routing. In *Proc. 7th Int. Workshop on Distributed Algorithms* (WDAG 93), Springer Verlag, LNCS 725, pp. 219-233, 1993.

[FJ88] G.N. Frederickson and R. Janardan. Designing networks with compact routing tables. *Algorithmica* **3**, pp. 171-190, 1988.

[FJ89] G.N. Frederickson and R. Janardan. Efficient message routing in planar networks. *SIAM Journal on Computing* **19**, pp. 843-857, 1989.

[FJ90] G.N. Frederickson and R. Janardan. Space-efficient message routing in c-decomposable networks. *SIAM Journal on Computing* **19**(1), pp. 164-181, 1990.

[FRU92] S. Felperin, P. Raghavan, E. Upfal. A theory of wormhole routing in parallel computers. In *Proc. of the 33rd IEEE Symp. on Foundations of Computer Science*, pp. 563-572, 1992.

[FS97] M. Flammini, C. Scheideler. Simple, efficient routing schemes for all-optical networks. In *9th Ann. ACM Symp. on Parallel Algorithms and Architectures*, pp. 170-179, 1997.

[FW78] S. Fortune, J. Wyllie. Parallelism in random access machines. In *10th Ann. ACM Symp. on Theory of Computing*, pp. 114-118, 1978.

[GG92] A. Greenberg, J. Goodman. Sharp approximate models of deflection routing in mesh networks. *IEEE Trans. on Communications* **41**, 1992.

[GH89] B. Goldberg, P. Hudak. Implementing functional programs on a hypercube multiprocessor. In *Proc. of the 4th Conf. on Hypercubes, Concurrent Computers and Applications*, pp. 489-503, 1989.

[GH92] A. Greenberg, B. Hajek. Deflection routing in hypercube networks. *IEEE Trans. on Communications*, pp. 1070-1081, 1992.

[Gi96] *The Mobile Communications Handbook*. J.D. Gibson (Ed.), CRC Press, 1996.

[GK80] M. Gerla, L. Kleinrock. Flow control: A comparative study. *IEEE Trans. on Communications* **28**(4), pp. 553-574, 1980.

[GLM+95] B. Ghosh, T. Leighton, B. Maggs, S. Muthukrishnan, C.G. Plaxton, R. Rajaraman, A.W. Richa, R.E. Tarjan, D. Zuckerman. Tight analysis of two local load balancing algorithms. In *Proc. of the 27th Ann. ACM Symp. on Theory of Computing*, pp. 548-557, 1995.

[GM94] B. Ghosh, S. Muthukrishnan. Dynamic load balancing on parallel and distributed networks by random matchings. In *Proc. of the 6th Ann. ACM Symp. on Parallel Algorithms and Architectures*, pp. 226-235, 1994.

[GM96] L.A. Goldberg, P.D. MacKenzie. Analysis of practical backoff protocols for contention resolution with multiple servers. In *Proc. of the 7th ACM-SIAM Symposium on Discrete Algorithms*, pp. 554-563, 1996.

[GMR94] L.A. Goldberg, Y. Matias, S. Rao. An optical simulation of shared memory. In *Proc. of the 6th Ann. ACM Symp. on Parallel Algorithms and Architectures*, pp. 257-267, 1994.

[Go78] L.M. Goldschlager. A unified approach to models of synchronous parallel machines. In *10th Ann. ACM Symp. on Theory of Computing*, pp. 89-94, 1978.

[GO93] R. Greenberg, H.C. Oh. Universal wormhole routing. Technical Report, University of Maryland, 1993.

[GP83] Z. Galil, W. Paul. An efficient general-purpose parallel computer. *J. of the ACM* **30**, pp. 360-387, 1983.

[GP95] C. Gavoille, S. Pérennès. Routing memory requirement for distributed networks. Research Report 95-37, I3S, Université de Nice-Sophia Antipolis, 06903 Sophia Antipolis, 1995.

[GPSS97] L. A. Goldberg, M. Paterson, A. Srinivasan, E. Sweedyk. Better approximation guarantees for job-shop scheduling. In *Proc. of the 8th Symp. on Discrete Algorithms*, pp. 599-608, 1997.

[Gr92] P.E. Green. *Fiber-Optic Communication Networks*. Prentice Hall, 1992.

[GVY93] N. Garg, V. Vazirani, M. Yannakakis. Approximate max-flow min-(multi)cut theorems and their applications. In *Proc. of the 25th Ann. ACM Symp. on Theory of Computing*, pp. 698-707, 1993

[GZ94a] O. Gerstel, S. Zaks. The virtual path layout problem in fast networks. In *Proc. of the 13th ACM Conf. on Principles of Distributed Computing*, pp. 235-243, 1994.

[GZ94b] O. Gerstel, S. Zaks. Path layout problem in ATM networks. In *Proc. of the 1st Colloquium on Structure, Information and Communication Complexity*, pp. 151-166, 1994.

[Ha91] B. Hajek. Bounds on evacuation time for deflection routing. *Distributed Computing* **5**, pp. 1-6, 1991.

[Ha95] L.A. Hall. Approximability of flow shop scheduling. In *Proc. of the 36th Ann. IEEE Symp. on Foundations of Computer Science*, pp. 82-91, 1995.

[HB94] K.T. Herley, G. Bilardi. Deterministic simulations of PRAMs on bounded-degree networks. *SIAM J. on Computing* **23**(2), pp. 276-292, 1994.

[HC+93] G.R. Hill, P.J. Chidgey et al. A transport network layer based on optical network elements. *J. of Lightwave Technology* **11**, pp. 667-677, 1993.

[He91] K.T. Herley. A note on the token distribution problem. *Information Processing Letters* **28**, pp. 329-334, 1991.

[HLM+90] S.H. Hosseini, B. Litow, M. Malkawi, J. McPherson, K. Vairavan. Analysis of a graph coloring based distributed load balancing algorithm. *J. of Parallel and Distributed Computing* **10**, pp. 160-166, 1990.

[HLR87] J. Håstad, T. Leighton, B. Rogoff. Analysis of backoff protocols for multiple access channels. In *Proc. of the 19th ACM Symposium on Theory of Computing*, pp. 241-253, 1987.

[Ho87] M. Hofri. *Probabilistic Analysis of Algorithms: On Computing Methodologies for Computer Algorithms Performance Evaluation*, Springer Verlag, 1987.

[HR90] T. Hagerup, C. Rüb. A guided tour of Chernoff bounds. *Information Processing Letters* **33**, pp. 305-308, 1989/90.

[HT93] A. Heinrich, S. Taylor. A parabolic theory of load balancing. Research Report Caltech-CS-TR-93-25, Caltech Scalable Concurrent Computation Lab, Pasadena, CA, March 1993.

[HW95] M. Harchol-Balter, D. Wolfe. Bounding delays in packet-routing networks. In *Proc. of the 27th ACM Symposium on Theory of Computing*, pp. 248-257, 1995.

[IEEE94] *IEEE Communications Magazine*, Dez. 1994.

[IK96] *Mobile Computing*, T. Imielinski, H. Korth (Eds.), Kluwer Academic Publishers, 1996.

[IL95] H. Izadpanah, C. Lin. Design, implementation, and system integration for a multi-gigabit WDM network. In *Applications of Photonic Technology*, edited by G.A. Lampropoulos, J. Chrostowski, and R.M. Measures, Plenum Press, 1995, pp. 19-24.

[JLT93] *Journal of Lightwave Technology* 11(5/6), May/June 1993

[JR92] J. Jájá, K.W. Ryu. Load balancing and routing on the hypercube and related networks. *J. of Parallel and Distributed Computing* 14, pp. 431-435, 1992.

[KFW94] R.K. Koeninger, M. Furtney, M. Walker. A shared MPP from Cray Research. *Digital Technical Journal* 6(2), pp. 8-21, 1994.

[KK77] L. Kleinrock, F. Kamoun. Hierarchical routing for large networks – performance evaluation and optimization. *Comput. Networks, ISDN Systems* 1, pp. 155-174, 1977.

[KK92] C. Kaklamanis, D. Krizanc. Optimal sorting on mesh-connected processor arrays. In *Proc. of the 4th Ann. ACM Symp. on Parallel Algorithms and Architectures*, pp. 50-59, 1992.

[KKP97] E. Kranakis, D. Krizanc, A. Pelc. Hop-congestion trade-offs for high-speed networks. *Int. Jornal of Foundations of Computer Science* 8, 1997.

[KKR93a] C. Kaklamanis, D. Krizanc, S. Rao. Hot-potato routing on processor arrays. In *Proc. of the 5th Ann. ACM Symp. on Parallel Algorithms and Architectures*, pp. 273-282, 1993.

[KKR93b] C. Kaklamanis, D. Krizanc, S. Rao. Universal emulations with sublogarithmic slowdown. In *Proc. of the 34th Ann. IEEE Symp. on Foundations of Computer Science*, pp. 341-350, 1993.

[KKT91] C. Kaklamanis, D. Krizanc, T. Tsantilas. Tigth bounds for oblivious routing in the hypercube. *Mathematical Systems Theory* 24, pp. 223-232, 1991.

[KL95] N. Kahale, T. Leighton. Greedy dynamic routing on arrays. In *Proc. of the 6th ACM-SIAM Symposium on Discrete Algorithms*, 1995.

[KLM93] R. Karp, M. Luby, F. Meyer auf der Heide. Efficient PRAM simulation on a distributed memory machine. Technical Report TR-RI-93-134, Paderborn University, Sept. 1993. To appear in Algorithmica. A preliminary version appeared in Proc. STOC'92, pp. 318-236.

[KLM+89] R. Koch, T. Leighton, B. Maggs, S. Rao, A. Rosenberg. Work-preserving emulations of fixed-connection networks. In *Proc. of the 21st Ann. ACM Symp. on Theory of Computing*, pp. 227-240, 1989.

[Kn73] D.E. Knuth. *The Art of Computer Programming, Volume 3: Sorting and Searching*. Addison-Wesley, 1973.

[Ko88] R.R. Koch. Increasing the size of a network by a constant factor can increase performance by more than a constant factor. In *Proc. of the 29th Ann. IEEE Symp. on Foundations of Computer Science*, pp. 221-230, 1988.

[KP92] M. Klugerman, C.G. Plaxton. Small depth counting networks. In *Proc. of the 24th Ann. ACM Symp. on Theory of Computing*, pp. 417-428, 1992.

[KPST90] P. Klein, S.A. Plotkin, C. Stein, É. Tardos. Faster approximation algorithms for the unit capacity concurrent flow problem with applications to routing and finding sparse cuts. In *Proc. of the 22nd Ann. ACM Symp. on Theory of Computing*, pp. 310-321, 1990.

[Kr91] D. Krizanc. Oblivious routing with limited buffer capacity. *J. of Computer and System Sciences* **43**, pp. 317-327, 1991.

[Kr96] D. Krizanc. Time-randomness trade-offs in parallel computation. *J. of Algorithms* **20**, pp. 1-19, 1996.

[KR94] C.P. Kruskal, K.J. Rappoport. Bandwidth-based lower bounds on slowdown for efficient emulations of fixed-connection networks. In *Proc. of the 6th Ann. ACM Symp. on Parallel Algorithms and Architectures*, pp. 132-139, 1994.

[KRS96] R. Kralovic, P. Ruzicka, D. Stefankovic. The complexity of shortest path and dilation bounded interval routing. Comenius University, Bratislava, 1996.

[KRT88] D. Krizanc, S. Rajasekaran, T. Tsantilas. Optimal routing algorithms for mesh-connected processor arrays. In *Aegean Workshop on Computing: VLSI Algorithms and Architectures*, Springer Verlag, LNCS 319, pp. 411-422, 1988.

[KS83] C.P. Kruskal, M. Snir. The performance of multistage interconnection networks for multiprocessors. *IEEE Trans. on Computers* **C-32**(12), pp. 1091-1098, 1983.

[KS85] R. Khatib, N. Santoro. Labelling and implicit routing in networks. *The Computer Journal* **28**, pp. 5-8, 1985.

[KSS94] M. Kaufmann, J. Sibeyn, T. Suel. Derandomizing algorithms for routing and sorting on meshes. In *Proc. of the 5th Ann. ACM-SIAM Symp. on Discrete Algorithms*, pp. 669-679, 1994.

[Ku87] M. Kunde. Optimal sorting on multi-dimensionally mesh-connected computers. In *Proc. of the 4th Symp. on Theoretical Aspects of Computer Science*, pp. 408-419, 1987.

[Ku88] M. Kunde. Routing and sorting on mesh-connected arrays. In *Aegean Workshop on Computing: VLSI Algorithms and Architectures*, Springer Velag, LNCS 319, pp. 423-433, 1988.

[Ku91] M. Kunde. Balanced routing: Towards the distance bound on grids. In *Proc. of the 3rd Ann. ACM Symp. on Parallel Algorithms and Architectures*, pp. 260-271, 1991.

[KU86] A. Karlin, E. Upfal. Parallel hashing – an efficient implementation of shared memory. In *Proc. of the 18th Ann. ACM Symp. on Theory of Computing*, pp. 160-168, 1986.

[KV86] S. Kapoor, P.M. Vaidya. Fast algorithms for convex quadratic programming and multicommodity flows. In *Proc. of the 18th Ann. ACM Symp. on Theory of Computing*, pp. 147-159, 1986.

[KZ88] R. Karp, Y. Zhang. A randomized parallel branch-and-bound procedure. In *Proc. of the 20th Ann. ACM Symp. on Theory of Computing*, pp. 290-300, 1988.

[LAD+92] C.E. Leiserson, Z.S. Abuhamdeh, D.C. Douglas, C.R. Feynman, B.C. Kuszmaul, M.A. St. Pierre, D.S. Wells, M.C. Wong, S.-W. Yang, R. Zak. The network architecture of the Connection Machine CM-5. In *Proc. of the 4th Ann. ACM Symp. on Parallel Algorithms and Architectures*, pp. 272-285, 1992.

[Le85] F.T. Leighton. Tigth bounds on the complexity of parallel sorting. *IEEE Trans. on Computers*, **C-34**(4), pp.344-354, 1985.

[Le92] F.T. Leighton. *Introduction to Parallel Algorithms and Architectures: Arrays · Trees · Hypercubes*. Morgan Kaufmann Publishers (San Mateo, CA, 1992)

[LK87] F.C.H. Lin, R.M. Keller. The gradient model load balancing method. *IEEE Trans. on Software Engineering* **13**(1), pp. 32-38, 1987.

[LM89] F.T. Leighton, B.M. Maggs. Expanders might be practical: Fast algorithms for routing around faults in multibutterflies. In *Proc. of the 30th Ann. ACM Symp. on Foundations of Computer Science*, pp. 384-389, 1989.

[LM92] F.T. Leighton, B.M. Maggs. Fast algorithms for routing around faults in multibutterflies and randomly-wired splitter networks. *IEEE Trans. on Computers* **41**(5), pp. 578-587, 1992.

[LMP+95] F.T. Leighton, F. Makedon, S. Plotkin, C. Stein, É. Tardos, S. Tragoudas. Fast approximation algorithms for multicommodity flow problems. *J. of Computer and System Sciences* **50**, pp. 228-243, 1995.

[LMR88] F.T. Leighton, B.M. Maggs, S.B. Rao. Universal packet routing algorithms. In *Proc. of the 29th Ann. Symp. on Foundations of Computer Science*, pp. 256-271, 1988.

[LMR94] F.T. Leighton, B.M. Maggs, S. Rao. Packet routing and job-shop scheduling in O(congestion + dilation) steps. *Combinatorica* **14**, pp. 167-186, 1994.

[LMR96] F.T. Leighton, B.M. Maggs, A.W. Richa. Fast algorithms for finding O(congestion+dilation) packet routing schedules. Technical Report CMU–CS–96–152, School of Computer Science, Carnegie Mellon University, Pittsburgh, PA, USA, 1996.

[LMRR94] F.T. Leighton, B.M. Maggs, A.G. Ranade, S.B. Rao. Randomized routing and sorting on fixed-connection networks. *Journal of Algorithms* **17**, pp. 157-205, 1994.

[LMS97] F.T. Leighton, B.M. Maggs, R. Sitaraman. On the fault tolerance of some popular bounded-degree networks. To appear in *SIAM Journal of Computing*.

[LMT89] F.T. Leighton, F. Makedon, I. Tollis. A $2N-2$ step algorithm for routing in an $N \times N$ mesh. In *Proc. of the 1st ACM Symp. on Parallel Algorithms and Architectures*, pp. 328-335, 1989.

[LPS88] A. Lubotzky, R. Phillips, R. Sarnak. Ramanujan graphs. *Combinatorica* **8**(3), pp. 261-277, 1988.

[LR88] F.T. Leighton, S. Rao. An approximate max-flow min-cut theorem for uniform multicommodity flow problems with applications to approximation algorithms. In *Proc. of the 29th Ann. IEEE Symp. on Foundations of Computer Science*, pp. 422-431, 1988.

[Ls85] C.E. Leiserson. Fat-Trees: Universal networks for hardware-efficient supercomputing. *IEEE Trans. on Computers* **C-34**, pp. 892-901, 1985.

[LT86] J. van Leeuwen, R.B. Tan. Routing with compact routing tables. In *The Book of L*, G. Rozenberg and A. Salomaa, eds. Springer Verlag, New York, pp. 259-273, 1986.

[LT87] J. van Leeuwen, R.B. Tan. Interval Routing. *The Computer Journal* **28**, pp. 298-307, 1987.

[LT79] R.J. Lipton, R.E. Tarjan. A separator theorem for planar graphs. *SIAM J. Appl. Math.* **36**, pp. 177-189, 1979.

[Lu96] R. Lüling. *Lastverteilungsverfahren zur effizienten Nutzung paralleler Systeme*. Ph.D. thesis, Paderborn University, 1996.

[LuM92] R. Lüling, B. Monien. Load balancing for distributed branch and bound algorithms. In *Proc. of the 6th Int. Parallel Processing Symposium*, pp. 543-549, 1992.

[LuM93] R. Lüling, B. Monien. A dynamic distributed load balancing algorithm with provable good performance. In *Proc. of the 5th Ann. ACM Symp. on Parallel Algorithms and Architectures*, pp. 164-172, 1993.

[Ma88] G.A. Margulis. Explicit group theoretical constructions of combinatorial schemes and their application to the design of expanders and superconcentrators. *Problems Inform. Transmission* **11**, pp. 39-46, 1988.

[Ma89] N. Maxemchuk. Comparison of deflection and store-and-forward techniques in the manhattan street and shuffle-exchange networks. In *Proc. of the IEEE INFOCOM*, pp. 800-809, 1989.

[MB76] R. Metcalfe, D. Boggs. Distributed packet switching for local computer networks. *Communications of the ACM* **19**, pp. 395-404, 1976.

[Me83] F. Meyer auf der Heide. Efficiency of universal parallel computers. *Acta Informatica* **19**, pp. 269-296, 1983.

[Me86] F. Meyer auf der Heide. Efficient simulations among several models of parallel computers. *SIAM J. on Computing* **15**, pp. 106-119, 1986.

[Mi86] G. Miller. Finding small simple cycle separators for 2-connected planar graphs. *J. of Computer System Sciences* **32**, pp. 265-279, 1986.

[Mi94] M. Mitzenmacher. Bounds on the greedy routing algorithm for array networks. In *Proc. of the 6th ACM Symposium on Parallel Algorithms and Architectures*, pp. 346-353, 1994.

[Mo94] M. Morgenstern. Existence and explicit constructions of $q + 1$ regular Ramanujan graphs for every prime power q. *Journal of Comb. Theory, Series B* **62**, pp. 44-62, 1994.

[MOW96] F. Meyer auf der Heide, B. Oesterdiekhoff, R. Wanka. Strongly adaptive token distribution. *Algorithmica* **15**, pp. 413-427, 1996.

[MS90] B. Monien, H. Sudborough. Embedding one interconnection network in another. *Computing* **7**, pp. 257-282, 1990.

[MS92] B.M. Maggs, R.K. Sitaraman. Simple algorithms for routing on butterfly networks with bounded queues. In *Proc. of the 24th Ann. ACM Symp. on Theory of Computing*, 1992.

[MS93] F. Matera, S. Settembre. All optical implementations of high capacity TDMA networks. *Fiber and Integrated Optics* **12**, pp. 173-186, 1993.

[MS95a] F. Meyer auf der Heide, C. Scheideler. Space-efficient routing in vertex-symmetric networks. In *Proc. of the 7th Ann. ACM Symp. on Parallel Algorithms and Architectures*, pp. 137-146, 1995.

[MS95b] F. Meyer auf der Heide, C. Scheideler. Routing with bounded buffers and hot-potato routing in vertex-symmetric networks. In *3rd European Symp. on Algorithms*, pp. 341-354, 1995.

[MS96a] F. Meyer auf der Heide, C. Scheideler. Deterministic routing with bounded buffers: Turning offline into online protocols. In *Proc. of the 37th Ann. IEEE Symp. on Foundations of Computer Science*, pp. 370-379, 1996.

[MS96b] F. Meyer auf der Heide, C. Scheideler. Communication in Parallel Systems. In *SOFSEM '96*, pp. 16-33, 1996.

[MSS96] F. Meyer auf der Heide, C. Scheideler, V. Stemann. Exploiting storage redundancy to speed up randomized shared memory simulations. *Theoretical Computer Science* **162**, July 1996.

[MSW95] F. Meyer auf der Heide, M. Storch, R. Wanka. Optimal trade-offs between size and slowdown for universal parallel networks. In *Proc. of the 7th Ann. ACM Symp. on Parallel Algorithms and Architectures*, pp. 119-128, 1995.

[MV84] K. Mehlhorn, U. Vishkin. Randomized and deterministic simulations of PRAMs by parallel machines with restricted granularity of parallel memories. *Acta Informatica* **21**, pp. 339-374, 1984.

[MV95] F. Meyer auf der Heide and B. Vöcking. A packet routing protocol for arbitrary networks. In *12th Symp. on Theoretical Aspects of Computer Science*, pp. 291-302, 1995.

[MV96] F. Meyer auf der Heide and B. Vöcking. Universal store-and-forward routing. Technical Report, Paderborn, 1996.

[MV97] B.M. Maggs, B. Vöcking. Improved routing and sorting on multibutterflies. In *29th Ann. ACM Symp. on Theory of Computing*, 517-530, 1997.

[MW89] F. Meyer auf der Heide, R. Wanka. Time-optimal simulations of networks by universal parallel computers. In *Proc. of the 6th Symp. on Theoretical Aspects of Computer Science*, pp. 120-131, 1989.

[MW95] F. Meyer auf der Heide, M. Westermann. Hot-potato routing on multi-dimensional tori. In *Int. Workshop on Distributed Algorithms*, 1995.

[MW96] F. Meyer auf der Heide , R. Wanka. Kommunikation in parallelen Rech-nernetzen (in German). In *Highlights aus der Informatik*, I. Wegener (editor), Springer Verlag, pp. 177-198, 1996.

[NM93] L.M. Ni, P.K. McKinley. A survey of wormhole routing techniques in direct networks. *IEEE Computer* **26**(2), pp. 62-76, 1993.

[NS79] D. Nassimi, S. Sahni. Bitonic sort on a mesh-connected parallel computer. *IEEE Trans. on Computers* **C-28**, pp. 2-7, 1979.

[NWD93] M.D. Noakes, D.A. Wallach, W.J. Dally. The J-Machine multicomputer: An architectural evaluation. In *Proc. of the 20th Int. Symp. on Computer Ar-chitecture*, pp. 224-235, 1993.

[NXG85] L.M. Ni, C. Xu, T.B. Gendreau. Distributed drafting algorithm for load balancing. *IEEE Trans. on Software Engineering* **11**(10), pp. 1153-1161, 1985.

[Or74] S.E. Orcutt. *Computer Organization and Algorithms for Very-High Speed Communications*. Ph.D. thesis, Department of Computer Science, Stanford Uni-versity, September 1974.

[OR97] R. Ostrovsky, Y. Rabani. Universal O(congestion + dilation + $\log^{1+\epsilon} N$) local control packet switching algorithms. In *29th Ann. ACM Symp. on Theory of Computing*, pp. 644-653, 1997.

[Pa90] I. Parberry. An optimal time bound for oblivious routing. *Algorithmica* **5**, pp. 243-250, 1990.

[Pa92] R.K. Pankaj. *Architectures for Linear Lightwave Networks*. Ph.D. thesis, MIT, 1992.

[Pe83] S. Personick. Review of fundamentals of optical fiber systems. *IEEE J. Se-lected Areas in Communications* **3**, pp. 373-380, April 1983.

[PG91] G.D. Pifarré, L. Gravano, S.A. Felperin, J.L.C. Sanz. Fully-adaptive min-imal deadlock-free packet routing in hypercubes, meshes, and other networks. In *Proc. of the 3rd Ann. ACM Symp. on Parallel Algorithms and Architectures*, pp. 278-290, 1991.

[PG95] R.K. Pankaj, R.G. Gallager. Wavelength requirements of all-optical net-works. *IEEE/ACM Trans. on Networking* **3**(3), pp. 269-280, 1995.

[Pi82] N. Pippenger. Telephone switching networks. In *Proc. Symposia in Applied Mathematics* **26**, American Mathematical Society, Providence, RI, pp. 101-133, 1982.

[Pi84] N. Pippenger. Parallel communication with limited buffers. In *Proc. of the 25th IEEE Symp. on Foundations of Computer Science*, pp. 127-136, 1984.

[Pl89] C.G. Plaxton. Load balancing, selection and sorting on the hypercube. In *Proc. of the 1st Ann. ACM Symp. on Parallel Algorithms and Architectures*, pp. 64-73, 1989.

[PS89] D. Peleg, A.A. Schäffer. Graph spanners. *Journal of Graph Theory* **13**, pp. 99-116, 1989.

[PS93] G.R. Pieris, G.H. Sasaki. A linear lightwave Beneš network. *IEEE/ACM Trans. on Networking* **1**, 1993.

[PS95] M. Paterson, A. Srinivasan. Contention resolution with bounded delay. In *Proc. of the 34th IEEE Symposium on Foundations of Computer Science*, 1995.

[PT93] S. Plotkin, É. Tardos. Improved bounds on the max-flow min-cut ratio for multi-commodity flows. In *Proc. of the 25th Ann. ACM Symp. on Theory of Computing*, pp. 691-697, 1993.

[PU87] D. Peleg and E. Upfal. The generalized packet routing problem. *Theoretical Computer Science* **53**, pp. 281-293, 1987.

[PU89a] D. Peleg and E. Upfal. The token distribution problem. *SIAM J. on Com-puting* **18**(2), pp. 229-243, 1989.

[PU89b] D. Peleg and E. Upfal. A tradeoff between size amd efficiency for routing tables. *Journal of the ACM* **36**, pp.510-530, 1989.

[Ra88] P. Raghavan. Probabilistic construction of deterministic algorithms: Approximating packing integer programs. *J. of Computer and System Sciences* **37**, pp. 130-143, 1988.

[Ra87] A.G. Ranade. How to emulate shared memory. In *Proc. of the 28th Ann. IEEE Symp. on Foundations of Computer Science*, pp. 185-194, 1987.

[Ra91] A.G. Ranade. How to emulate shared memory. *J. of Computer and System Sciences* **42**, pp. 307-326, 1991.

[Ra93] R. Ramaswami. Multi-wavelength lightwave networks for computer communication. *IEEE Communications Magazine* **31**, pp. 78-88, 1993.

[Ro78] L.G. Roberts. The evolution of packet switching. *Proceedings of the IEEE* **66**(11), pp. 1307-1313, 1978.

[RS95] R. Ramaswami, K.N. Sivarajan. Routing and wavelength assignment in all-optical networks. *IEEE/ACM Trans. on Networking* **3**(5), pp. 489-500, 1995.

[RSW94] A. Ranade, S. Schleimer, D.S. Wilkerson. Nearly tight bounds for wormhole routing. In *Proc. of the 35th Ann. IEEE Symp. on Foundations of Computer Science*, pp. 347-355, 1994.

[RT96] Y. Rabani, É. Tardos. Distributed packet switching in arbitrary networks. In *28th Ann. ACM Symp. on Theory of Computing*, pp. 366-375, 1996.

[RU94] P. Raghavan, E. Upfal. Efficient routing in all-optical networks. In *Proc. of the 26th Ann. ACM Symp. on Theory of Computing*, 1994.

[RU95] P. Raghavan, E. Upfal. Stochastic contention resolution with short delays. In *Proc. of the 27th ACM Symposium on Theory of Computing*, pp. 229-237, 1995.

[SAVP94] J. Subias, R. Alonso, F. Villuendas, J. Pelayo. Wavelength selective optical fiber couplers based on longitudinal Fabry-Perot structures. *J. of Lightwave Technology* **12**, pp. 1129-1135, 1994.

[Sch90] E.J. Schwabe. On the computational equivalence of hypercube-derived networks. In *Proc. of the 2nd Ann. ACM Symp. on Parallel Algorithms and Architectures*, pp. 388-397, 1990.

[SJ95] L. Schwiebert, D.N. Jayashima. A necessary and sufficient condition for deadlock-free wormhole routing. In *Proc. of the 7th ACM Symp. on Parallel Algorithms and Architectures*, pp. 175-184, 1995.

[SM93] M. Settembre, F. Matera. All optical implementations of high capacity TDMA networks. *Fiber and Integrated Optics* **12**, pp. 173-186, 1993.

[SS79] W.J. Savitch, M. Stimson. Time bounded random access machines with parallel processing. *Journal of the ACM*, pp. 103-118, 1979.

[SS86] C.P. Schnorr, A. Shamir. On optimal sorting algorithm for mesh-connected computers. In *Proc. of the 18th Ann. ACM Symp. on Theory of Computing*, pp. 255-263, 1986.

[SS90] J.P. Schmidt, A. Siegel. The spatial complexity of oblivious k-probe hash functions. *SIAM J. on Computing* **19**(5), pp. 775-786, 1990.

[SS94] R. Subramanian , I.D. Scherson. An analysis of diffusive load balancing. In *Proc. of the 6th Ann. ACM Symp. on Parallel Algorithms and Architectures*, pp. 220-225, 1994.

[SSS95] J.P. Schmidt, A. Siegel, A. Srinivasan. Chernoff-Hoeffding bounds for applications with limited independence. *SIAM J. on Discrete Mathematics* **8**(2), pp. 223-250, 1995.

[SSW94] D.B. Shmoys, C. Stein, J. Wein. Improved approximation algorithms for shop scheduling problems. *SIAM Journal of Computing* **23**(3), pp. 617-632, 1994.

[ST97] A. Srinivasan, C.-P. Teo. A constant-factor approximation algorithm for packet routing, and balancing local vs. global criteria. In *29th Ann. ACM Symp. on Theory of Computing*, 1997.

[Su94] T. Suel. Improved bounds for routing and sorting on multi-dimensional meshes. In *Proc. of the 6th Ann. ACM Symp. on Parallel Algorithms and Architectures*, pp. 26-35, 1994.

[SV93] O. Sýkora, I. Vrťo. Edge seperators for graphs of bounded genus with applications. *Theoretical Computer Science* **112**, pp. 419-429, 1993.

[SV96] C. Scheideler, B. Vöcking. Universal continuous routing strategies. In *Proc. of the 8th Ann. ACM Symp. on Parallel Algorithms and Architectures*, pp. 142-151, 1996.

[TK77] C.D. Thompson, H.T. Kung. Sorting on a mesh-connected parallel computer. *Communications of the ACM* **20**, pp. 263-271, 1977.

[TW91] A. Trew, G. Wilson. *Past, Present, Parallel – A Survey of Available Parallel Computing Systems*, Springer Verlag, 1991.

[Up82] E. Upfal. Efficient schemes for parallel communication. In *Proc. of the 1st Ann. ACM Symp. on Principles of Distributed Computing*, pp. 241-250, 1982.

[Up89] E. Upfal. An $O(\log N)$ deterministic packet routing scheme. In *Proc. of the 21st Ann. ACM Symp. on Theory of Computing*, pp. 241-250, 1989.

[Up92] E. Upfal. An $O(\log N)$ deterministic packet routing scheme. *Journal of the ACM* **39**, pp. 55-70, 1992.

[UW87] E. Upfal, A. Wigderson. How to share memory in a distributed system. *Journal of the ACM* **34**, pp. 116-127, 1987.

[Va82] L.G. Valiant. A scheme for fast parallel communication. *SIAM J. on Computing* **11**(2), pp. 350-361, 1982.

[Va90] L.G. Valiant. A bridging model for parallel computation. *Communications of the ACM* **33**(8), pp. 103–111, 1990.

[Va94] V.V. Vazirani. A theory of alternating paths and blossoms for proving correctness of the $O(\sqrt{|V|}|E|)$ general graph maximum matching algorithm. *Combinatorica* **14**(1), pp. 71-109, 1994.

[Wa68] A. Waksman. A permutation network. *Journal of the ACM* **15**(1), pp. 159-163, 1968.

[Wi91] R.D. Williams. Performance of dynamic load balancing algorithms for unstructured mesh calculations. *Concurrency: Practice and Experience* **3**(5), pp. 457-481, 1991.

[ZA95] Z. Zhang, A.S. Acampora. A heuristic wavelength assignment algorithm for multihop WDM networks with wavelength routing and wavelength re-use. *IEEE/ACM Trans. on Networking* **3**(3), pp. 281-288, 1995.

Index

Springer
and the
environment

At Springer we firmly believe that an
international science publisher has a
special obligation to the environment,
and our corporate policies consistently
reflect this conviction.
We also expect our business partners –
paper mills, printers, packaging
manufacturers, etc. – to commit
themselves to using materials and
production processes that do not harm
the environment. The paper in this
book is made from low- or no-chlorine
pulp and is acid free, in conformance
with international standards for paper
permanency.

Lecture Notes in Computer Science

For information about Vols. 1–1332

please contact your bookseller or Springer-Verlag